Processing and Conveying Equipment Design

Prabhakar S. Phirke
Ex-Head of the Department
Agricultural Process Engineering
Dr. Panjabrao Krishi Vidyapeeth
Akola (Maharashtra)

S.C.B. Siripurapu
Ex-Principal
College of Agril. Engineering and Technology
Gujarat Agricultural University
Junagadh (Gujarat)

2009

Jain Brothers (New Delhi)
16/873, East Park Road, Karol Bagh,
New Delhi - 110 005
Ph. : (011) 23618426, 45097398
E-mail : jain_bros@hotmail.com
Website : www.thejainbrothers.com

Published by :

Shri Sunil Kumar Jain
for
Jain Brothers (New Delhi)
16/873, East Park Road,
New Delhi – 110 005

Branch Office at **:**

- Fatima Nagar, Opp. Andhra Bank,
  KAZIPET - 506004
  Ph. : 0870-2459617

First Published : 2004
Second edition : 2009

ISBN : 978-81-8360-109-2

To
Sanjivani
Prabhakar
Phirke

Preface to the Second Edition

It is a great pleasure to present the second edition of this book to the students and concerned teachers. In presenting this second edition, the authors like to thank to many friends from the country that have kindly made suggestions for the different parts of this textbook. Special thanks are due to many readers who have made positive suggestions. Resulting small changes can be observed throughout the text in terms of uniform use of variables, their nomenclature and precise solutions of the worked examples.

Date 15-6-2009

Prabhakar S. Phirke
S.C.B. Siripurapu

Preface to the First Edition

A subject ***Process Equipment Design*** is recently introduced at undergraduate level of B. Tech. (Agril. Engg.) degree programme (1998-99). As such not a single consolidated book is available on the subject where variety of machines are included in the curriculum of Indian Agricultural Universities in a common course on the pattern given by Indian Council of Agricultural Research, New Delhi. This book is in an attempt to put together different types of processing and conveying equipments required in agricultural sector, particulary, those covered in the said curriculum.

This book is mainly intended for the Agricultural Engineering students who have already studied the prerequisite subjects such as Theory of Machines, Strength of Materials. Thermodynamics, Heat and Mass Transfer and Machine Drawing in their earlier years of study.

Basic design principles and parameters are dealt in Chapter 1 and selection of material for equipment in Chapter 2. Different process equipments (pressure and storage vessels, heat exchanges, evaporators and their supports) are covered in Chapter 3 through Chapter 7. Chapters 8 and 9 present the design of grain dryers and spray dryers, respectively. Design of grain milling equipments is presented in Chapter 10. The next four chapters are devoted to the conveying machines *viz.* belt conveyors, bucket elevators, screw conveyers and pneumatic conveyers where, the general description, operating principle and design are enumerated. Any designer can fail in his profession if economy is neglected. The way and methods of achieving the economy in sizing the equipments are discussed in Chapter 15 with preliminary approaches of optimization.

Most of the equipments are discussed prior to their design on different points such as their types, working principle, constructional features, applications etc. The design procedures are outlined such that an independent unit or its part by part design can be done with ease in understanding. Likewise, small and full length examples are solved unitwise and also complete unit together is also dealt in a simple way. This will make the student to understand the subject to a depth. All the formulae, data, constants and examples are presented in the SI system of units which is accepted worldwide.

About one hundred illustrations through line diagrams are included for clarity and clear understanding. The relevent literature is honestly acknowledged through the references but there may be inevitably crept errors or omissions in this book. If such errors/omissions are pointed out, it will be suitably corrected.

I am grateful to Dr. N.G. Bhole, Ex-Professor and Head PHTC, Indian Institute of Technology, Kharagpur and Dr. S.V. Supe, Ex-Head, Department of Agricultural Extension who inspired me and suggested to write a textbook revising my lecture notes on the subject.

P. S. Phirke

Table of Contents

1

Design Principles and Parameters

1.1 Introduction

Design, in general, is a creative activity or synthesis to put the ideas together for achievement of the desired objective. In this era of change, either new/ modern plants are being established or the old ones are modernized by alteration of unit operations and equipments, in turn. This requires the knowledge of unit operations in the processing of raw to finished products, thermodynamics, heat transfer, process plant design and equipment design. The shape and size of plant is governed by the process to be followed and associated machinery, vessels and equipments. With a specific objective in mind, the designer starts thinking about a specific problem, puts forth some ideas and constraints and by proper evaluation comes to the conclusion to achieve the target. For example, for a food process engineer, either a new food product or a new stage in a production process can be the target for design.

At different stages of unit operations in a process, many equipments are required for handling, processing and storage of food products. There may be a particular equipment for a specific task/ purpose in a processing plant, or with little modification, it can be put to the another function with plant itself. There should be provision in processing equipment to take such allied functions and allowance for accidental overloading. It should perform assured function. The equipment reliability depends on many factors such as (*a*) strength and rigidity of components. (*b*) reliable fabrication, (*c*) material of construction, (*d*) process conditions, (*e*) ease of operation and control, (*f*) easy repairs and maintenance, (*g*) safe working and (*h*) economic considerations.

1.2 General design methodology

Design is an iterative process as shown in Fig. 1.1, the designer being aware of more possible designs and constraints thereon for number of iterations. He should constantly seek the new ideas and data with the advance of the design and evaluate the requisite design solution. Although the design procedure can not be standardized, it should pass through the following stages.

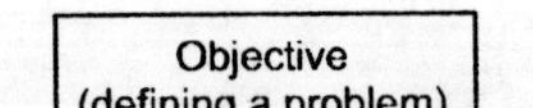

Fig. 1.1. *The general iterative design process*

(1) Defining a problem (objective)

(2) Analysis and constraints

(3) Generation of possible designs

(4) Selection of construction materials and process condition.

(5) Evaluation and optimization of the design

(6) Equipment specifications and fabrication drawings.

The first and foremost stage is to understand and define a problem or objective to fulfil a specific need. It may arise as a commercial opportunity or a new idea moulded in the new form, market survey, imagination or invention. The designer should prepare a clear and complete statement of the needs in detail as far as possible and formulate secondary or sub-objectives with awareness of the constraints. The idea or methodology must be analysed both qualitatively and quantitatively to put it for satisfactory performance during working. A comprehensive and specific need defines the external constraints within which the other designers must work. The needs and wants should be clearly separated out because the wants are those parts of initial specification which may be thought desirable but can be relaxed if required as the design progresses. For instance, a particular product may be costly and difficult to manufacture. In this case, some relaxation in specifications may evolve a product that is cheaper and saleable too. The designer must have sufficient data regarding possible processes, physical and chemical properties of the product, equipment performance etc. which may be obtained from national standards, design manuals published by organizations and books.

Considering the process principles and mechanics of the equipments, they should be designed mathematically and technically to generate few possible solutions to the problem concerned. This calls for the experience of the designer which enables him to select well-tried and tested methodology and technology rather than exciting but untried novel design. The design of vessels or equipments may be required in three-fold: (*a*) modification and addition to existing plant, (*b*) repetition of existing designs with minor changes and enhancing the production capacity and (*c*) new processes from innovation or pilot plant to a commercial process. To work the design for any of the above purpose, the first step is to draw a rough sketch and block diagram along with a flow chart showing the unit operations and milestones in a process clearly stating the functions and constraints at each stage. A clear statement of process conditions may enable to deduce the bulk of design through analytical approximations. The selection of construction material for the equipments can not be separated from design process since the allowable stresses, imposed stresses and physical properties of materials will affect the possible designs.

A check and cross-check to all the mathematical calculations should be made and should be evaluated both for technical and economical feasibility. Moreover, the results should also be evaluated whether they are practicable and are matching to the past experiences in respect of controllable parameters. The maximum or minimum final results should be obtained through iterations till it gives a satisfactory solution. The designer should use some optimization techniques to arrive at the final result such as mathematical modelling, analytical methods, statistical research methods, linear programming etc. If a new technique or equipment is selected, its fabrication and testing is necessary before finalization of the design. The sketches of designed parts or equipments should be associated with computations so that it is easy to specify the materials selected, fits, limits and tolerances, surface and strength requirements, finishes etc. This can also facilitate to draw fabrication assembly drawings with ease and quickly. While designing and drawing an attention should be paid for dismantling, assembly and maintenance which may be as simple as possible to reduce the cost of repairs and maintenance.

1.3 Design codes and standards

Use of codes and standards in the design has started in 1841 when Whitworth introduced first standard screw thread so as to have the interchangeability amongst various manufacturers. Although the two terms are interchangeably used, code means the code of practice *i.e.*, a particular operating procedure or the recommended design whereas standard is used for particular size or compositions etc. In practice, adoption of the standard design saves time and money ensuring the confirmation between various projects due to the reason that most of the engineering designs are routine and repetitive in nature. Equipment manufactures work to standards to produce the standardized designs and size ranges for commonly used machines or elements such as pipes and pipe fittings, electric motors, pumps etc. Clearly, it is more economical to produce limited range of standard size rather than to go each job as a special work. The stock of spares is considerably reduced due to adoption of standards. For the designer, specifying a particular standard equipment, automatically tells many dimensions or performances of a specific item. Besides this, the standards impose constraints on the designer. The nearest rounded computed size may not be optimum but may be economical due to standardization. The standard design methods must, of their nature, be historical and do not necessarily incorporate the latest techniques. However, this difficulty can be overcome by the revision of prevailing codes and standards to introduce new methodology or techniques.

1.4 Design variables

Some operations are performed on the intake material streams to produce the desired outlet streams in processing plant. The operation of unit can be mathematically modelled in the form of equations. In the design unit the flow of information can be replaced by flow of material in processing plants. The information flows are the values of variables involved in design *e.g.*, temperature, pressure, flowrate of streams along with enthalpies and composition of streams etc. The temperature, pressure and composition of stream are intensive variables, independent of the flow of material quantitatively. The design constraints will impose some limitations on the possible values that these variables can take. Some of the variable ranges can be defined directly from the process specifications. The values of other design variables can be computed by using the constrained design relationships. Some of the design relationships will be in the form of explicit mathematical equations or design equations, for example, those arising from thermodynamic and heat transfer relationships, mass and energy balances and parameters of equipment performance. Few other approximate relationships can be obtained from the codes, standards, and safety considerations.

Number of variables involved in the design minus the number of design equations is the number of degrees of freedom. The number of variables in the system is analogous to the number of variables in a set of simultaneous mathematical equations and the number of relationships is analogous to the number of equations. In the design, the designer must define a problem clearly to specify a certain number of variables and then he can manipulate the variables to achieve the best design.

2

Selection of Materials

2.1 Introduction

The first and foremost requirement in designing the process equipments is the type and specifications of the material of their construction. In spite of many factors influencing the material selection, three important primary factors are mechanical strength, corrosion resistance and cost of the material and many times a wise compromise is necessary between these three factors. However, the process designer must consider the process conditions required such as temperature, pressure, type of loading, corrosion resistance, and specific effects on material due to process product. The mechanical properties and cost effectiveness against service life which satisfy the process as well as mechanical requirements should be the basis for specifying the material. Effects of fabrication techniques such as cutting, forming, welding, heat treatment, coating etc. should also be considered.

2.2 Material characteristics

The following characteristics of the construction material should be considered while selection.

(*a*) Mechanical properties such as strength, rigidity, elasticity, hardness, ductility, yield stress, fatigue and creep etc.

(*b*) Effect of working temperature on properties of the material.

(*c*) Corrosion resistance.

(*d*) Special properties like thermal conductivity, electrical conductivity and magnetic properties.

(*e*) Fabrication techniques required like forming, casting, welding etc.

(*f*) Commercial availability in the market.

(*g*) Cost effectiveness against service life.

2.2.1 Mechanical properties

Following are some of the important mechanical properties of the materials commonly used in the construction of process equipment.

(1) *Tensile strength*: It is the maximum stress which the material can withstand under tensile load, measured by a standard tensile test. Tensile strength (stress) forms a measure of the basic strength of material and is a function of Young's modulus of elasticity.

(2) *Stiffness*: It is the ability of material to resist deformation under bending and buckling. It is also called as rigidity which is a function of modulus of elasticity and second moment of area.

(3) *Toughness*: It is associated with tensile strength and is the ability of material to absorb energy in deformation in the plastic range. Total area under stress-strain curve gives its measure. In case of materials such as steel, aluminium and copper they locally yield at crack tip and stop the further propagation. In the materials like glass and cast iron local yielding does not occur and the materials are brittle.

(4) *Hardness*: The surface hardness is an indication of the ability of material to resist wear, may be scratching or indentation. This property must be considered while designing the equipments handling the abrasive solids or liquids with solids causing surface erosion.

(5) *Fatigue*: It represents the failure likely to occur when the equipment is working under cyclic loading e.g., rotating machines likely compressors and pumps. Initial small cracks are progressively enlarged under cyclic load or pressure.

(6) *Creep*: The creep is a gradual and progressive deformation of the material under steady tensile stress over prolonged period of time. The creep strength of material is a measure of stress to cause rupture in $(10)^5$ hours at a particular temperature.

2.2.2 Effect of working temperature on material properties

High temperature: With the increase in working temperature, the tensile strength and elastic modulus of materials decreases along with the creep strength and surface stability *e.g.*, low carbon mild steel strength is 450 N/mm^2 at 25 °C which reduces to 210 N/mm^2 at 500 °C with corresponding decrease in modulus of elasticity from $2(10)^5$ to $1{\cdot}5(10)^5$ N/mm^2. On the other hand, stainless steels can withstand high temperatures and least affecting the elastic modulus. The materials having higher coefficient of thermal expansion, expand with rise in temperature. These facts should be taken in to consideration for the design of an equipment component to be operated at higher temperatures.

Carbon steels are susceptible to heavy corrosion. However, addition of chromium (Cr), molybdenum (Mo), manganese (Mn) and vanadium (V) add to the resistance to scaling and oxidation and improves strength at higher temperatures. For instance, carbon steel can withstand temperature up to 400 °C, but addition of 0·5% Mo elevates it to 500 °C; 5% Cr and 0·5% Mo alloy works up to 620 °C but 15% Cr and 20% Ni elevates to 1100 °C and so forth. Furnace and heater tubes and other furnace parts require creep-rupture strength, internal stability and resistance to scaling. Austenitic stainless steels (25 Cr 12 Ni and 25 Cr 20 Ni) offer the best combination for temperature of 850 °C and above. In addition, stainless steels have corrosion resistance ability and adequate strength at high temperatures.

Low temperature: Notch-toughness is an important property of metals at low temperature which is the ability to deform plastically at stress concentrations and resistance to brittle cracking at notches under multiaxial stresses. Ferric steels show a sudden decrease in notch-toughness but nonferrous metals exhibit good correlation in ductility even at very low temperatures. Silicon-killed boiler quality carbon steel can be used for fabrication of low temperature process vessels to about –60 °C below 12 mm thickness under stress relieved condition. Nickel (3·57) steel can work for low temperature service vessels greater than –100 °C whereas 9% Ni-steel can be used for greater than –190°C. The austenitic stabilized steel (18 Cr 8 N; low C) can be successfully used for such service in welded construction.

2.2.3 Corrosion resistance

Composition of the construction materials is of vital importance from the point of view of process vessel safety, personal safety and product purity and must be considered seriously in food processing vessels. However, proper selection should be a wise compromise between the mechanical properties, corrosion resistance and the cost involved. Corrosion may be oxidative and non-oxidative; the former being most common including ordinary atmospheric corrosion and chemical attack by the product to be processed. In non-oxidative corrosion, a metal of vessel or equipment parts is dissolved or disintegrated but is not chemically oxidized. Metallic corrosion is essentially an electro non-corroding or passive element; electrolyte: corroding fluid normally conducting medium and the completion of electrical circuit through the material.

The oxidative corrosion rate is affected by many factors such as heat, light, surface homogeneity and presence or absence of polarizers. Various patterns of attack are produced due to variation in these factors namely, uniform corrosion, stress corrosion, dezincification and erosion corrosion. Uniform corrosion is least intense in its action and can be easily predicted. Corrosion allowance based on the annual decay rate and service life is provided as an extra thickness to safeguard this corrosion. If two dissimilar metals are in contact in presence of the electrolyte, the anodic metal will be corroded fast since the other metal would act as cathode causing galvanic corrosion. Stainless steel in oxidizing environment forms a natural protective film under certain conditions. Pitting corrosion occurs at small local areas without general corrosion resulting in pitting leading to wall decay. The intergranular corrosion is preferential at the grain (crystal) without general loss of the material. This can be avoided by using low carbon grades or grades stabilized by the addition of titanium or nibobium.

When the material is under stress, the corrosion rate and the form of attack may change e.g., simple corrosion stress, corrosion cracking and corrosion fatigue. Undue tensile stress development due to cold working or improper welding may cause cracking of the metals. These cracks grow rapidly and cause premature brittle failure. Caustic embrittlement of iron and steel and season cracking of brass are the examples of stress corrosion. The corrosion fatigue occurs when the metal is subjected to repeated reversals of stress in a corrosive medium. Dezincification is the process of leaching zinc from brass by the corrosive environment. Graphitization is a similar phenomenon where iron is corroded from cast iron leaving a weak graphite behind. Erosion-corrosion is a combination of mechanical wear (erosion) and corrosive nature. Pump impellers, body, agitators, pipes fittings etc. are usually affected by erosion and corrosion. To prevent this, wear resistance metals should be recommended for fabrication of such parts or surface coatings may be provided e.g., at the inlets of exchanger tubes plastic inserts are used.

The process environment to which the material will be exposed should be clearly defined such as pressure, temperature, pH value of fluid, presence of stress impurities (stress corrosion), amount of aeration (of differential oxidation cells), stream flow velocity and agitation (erosion corrosion) and heat transfer rates (differential temperatures). Corrosion resistant metals should be selected for a particular range of chemicals to be processed from the standard charts available in standards. A slight change in the process conditions or presence of unexpected impurities can markedly change the rate of attack or the nature of corrosion.

2.2.4. Cost of material

While selecting the construction material for vessels or other parts of the process equipments, its cost must be taken into account. The quantity of a material used depends on the density and design stress (strength) of the material. One of the ways to compare the material cost is by the following expression.

$$CR = \frac{c\rho}{\sigma_d}$$

where, CR is the cost rating used for comparison, c is the cost of material per unit mass, ρ is the density and σ_d is the design stress. Materials with a relatively high design stress like, stainless steels and low alloy steel can be used more cost effectively than low carbon steels. In addition to this, the cost of fabrication and annual cost of material must be considered for the selection of material. The annual cost is a function of its predicted life calculated from the corrosion rate and the purchased cost of equipment. Sometimes, it may be economical to use highly corrosive cheaper material and frequent replacement than high corrosion resistance material, under particular situation. If thick plates are required for fabrication for structural strength like pressure vessels, provision of clad material may reduce the cost of equipment substantially.

2.3 Commonly used materials

The materials of construction are mainly classified in two groups i.e., metals and non-metals. The metals are ferrous and non-ferrous type. Appendix A enumerates important properties of the materials.

2.3.1 Ferrous metals

Iron and steels are included in this group which have good strength, ease of forming of machining and are widely available. Their alloys with other metals give a wide range of properties under various adoptions. They are classified as wrought iron, cast iron, and steels depending on the carbon percentage content. Wrought iron is pure iron with low carbon and manganese content. It has good corrosion resistance in atmospheric conditions, in alkaline solutions and concentrated cold nitric and sulphuric acids. It has the high temperature applications like exhaust lines, heating coils, steam supply lines and smoke stacks. Cast iron contains carbon between 2·5 and 4·5%, easy to form the complicated shapes and cheaply available. It is of two types namely, gray cast iron in which graphite crystals are freely distributed throughout the mass, and white cast iron where iron and carbon are present in chemical combination. The former is not corroded at any temperature by $< 30\%$ concentrated alkalies and is used in special piping, pumps, valves and concentration vessels. It is particularly used where vessels of complicated shape are to be fabricated and where resistance to corrosion and abrasion is desired. White cast iron is highly brittle and is used either by heat treatment or by addition of small amount of calcium silicate or both to impart tensile and shear strength. It is used for pipe fittings and high pressure vessels.

Steel is a iron alloyed with carbon from 0·05 to 2·0% with small proportions of silicon, manganese, phosphorus and sulphur and are called as plain carbon steel classified under three groups *i.e.*, low, medium and high carbon steels. Low carbon steel contains carbon from 0·05 to 0·30% also called as mild steel. It is a versatile metal having good strength and ductility, can be easily rolled formed or forged. Machining and welding operations are easy and hence, are used widely in pressure vessels, machine components, pipes and fittings and standard structural sections. While using for corrosive product, the surface may be coated with either paint, metal spraying or other suitable lining. Low carbon steel is the most economical where the rate of corrosion is slow or negligible. Medium carbon steel has 0·3 to 0·5% carbon content which imparts higher hardness, strength and fatigue resistance. It is mostly used for bolts, gears, springs, and shafts etc. The higher carbon content from 0·5 to 2·0% in steel refer to high carbon steel used basically for dies and cutting tools since it is hardest and less ductile among carbon steels.

When the other elements such as Cr, Mo and Ni are added to steels, they have an increased strength and certain desired properties are obtained, namely, creep resistance, toughness, corrosion resistance, ability to withstand high working temperature and so forth. They can be divided in two groups depending on the proportion of alloying the other elements *i.e.*, low alloy steels with 10% and high alloy steels having alloying elements

>10%. Low alloy steels have thrice the yield stress values than that of ordinary mild steel of structural or boiler quality, resulting in a considerable weight saving for equipments and hence, the economy of material. Alloying Ni increase notch toughness of the low alloy steels. When corrosion resistance in power industries and nuclear installations is desired, high alloy steels are used. They are of two types: ferric and austenitic stainless steels contain chromium as 16-25% having good strength, ductility and high strain hardening coefficient *i.e.*, their hardness and tensile strength increases considerably on cold working. These steels are suitable under high temperature working conditions. Clad steel is a corrosion resistance composite product with carbon steel as a base material and 2 to 5 mm thick cladding of stainless steel which is homogeneously bonded by rolling. The mechanical properties of clad steel are the same as those of the backing steel and corrosion resistance is equal to that of the cladding materials.

2.3.2 Non-ferrous metals

In special industrial applications of food industry, dairy and chemical industries non-ferrous metals find particular place, the most common being, aluminium, copper, nickel and their alloys. Platinum, silver and lead are used either alone or as liners. Titanium, zirconium and nibobium metals have increasing applications.

2.3.2.1 Aluminium and its alloys

These materials are readily available, light, easy to fabricate and relatively inexpensive. Pure aluminium has good thermal conductivity and is weak in strength but offers good corrosion resistance in chemical environment due to formation of inert hydrated aluminium oxide and its tendency to adhere on the surface. Copper, silicon, zinc, magnesium, manganese and nickel are the most common elements used for alloying with aluminium. The ductility of all these alloys is maintained at low temperature to the tune of –200 °C. However, at elevated temperatures the mechanical strength of these alloys is reduced and hence, they are used for working temperatures of less than 150 °C. Aluminium alloys find wide applications in process vessels and systems in dairy and food processing, refrigeration industries, storage vessels and heat exchangers.

2.3.2.2 Copper and its alloys

Pure copper has good malleability, ductility and high electrical and thermal conductivity. It has fair mechanical strength and can be easily joined by soldering, brazing or welding. Copper forms a protective coating in atmospheric moisture or oxygen, strong alkalies and organic solvents due to formation of copper oxides. Copper exhibits poor creep strength above 150 °C and hence, used in low temperature services only in food processing plants, evaporators, heat exchangers and in manufacture and recovery of organic solvents. Zinc to the extent of 45% with copper is known as brass which exhibit less corrosion resistance than copper itself. They fail by stress corrosion cracking and have limited temperature. Bronzes are copper-tin alloys containing 5% tin for wrought work and 70% for castings, widely used in tubing, pipes and fittings, values and pumps. Copper-aluminium containing 14% aluminium, known as aluminium bronzes, are resistant to oxidation, corrosion and much stronger than pure copper, especially above 750 °C. They are used for pump castings, value seat and condenser tubes. Aluminium bronze with Fe and Ni has good ductility, strength and corrosion resistance and extensively used in the fabrication of condensers, tube sheets up to 350°C working temperatures.

2.3.2.3 Nickel and its alloys

These are very expensive metals and hence, their use is limited for cladding of the process vessels and in heat exchanger tubes, tube sheets and small thin- walled vessels. They can handle a variety of fluids both at low and elevated temperatures where it is essential to prevent the contamination of contained fluid. Monel, inconel and incoloy are the main nickel, alloys widely used. Monel alloy contains 67 Ni and 30 Cu which has high strength, ductility and excellent corrosion resistance. They are mostly used for cladding of carbon steels or low-alloy steels. Inconel (16 Cr 8 Fe 1 Mn) have good strength and corrosion resistance at low and elevated temperatures. They are widely used in dairy and food processing vessels, alkaline mediums and organic acids. Incoloy (21 Cr 44 Fe 1·5 Mn 33·5 Ni) has similar properties to that of inconel and essentially is a high temperature alloy with high creep- rupture strength and structural stability even up to 900 °C.

2.4 Non-metals

Number of materials used for fabrication of the process vessels are equipments other than metals as described above. They are plastics, ceramic, rubber, glass, carbon and graphite etc.

Plastics: Plastics are being increasingly used in the construction of processing equipments. As compared with the metals, plastics have low mechanical strength and are used for moderate pressure and temperature. They have excellent resistance to weak mineral acids and salt solutions and are not corroded under ambient conditions. Plastics have low weight, good electrical and thermal resistance (insulations) pleasing appearance and ease of fabrication. They are divided in two broad classes: thermoplastics which soften with increase in temperature *e.g.*, polyvinyl chloride (PVC) and polyethylene and second, thermosetting plastics which have a rigid, cross-linked structure e.g., polyester and epoxy resin. PVC, polythene, polypropelene, and cellulose acetate are the widely used thermoplastics available commercially as films, sheets, rods and pipes and find their use in pipes, fittings, gaskets, fans, ventilation systems, and storage systems. The thermosetting plastics are hard and brittle and are present in combination with filters such as cotton flock, wood flock or asbestos etc. Thermosetting plastics with glass fibre reinforcement are used in manufacture of pipes, fittings, ducts and vessels. They can be moulded to any shape and are also used as lining and laminates.

Rubber : It is used particularly as lining material to the vessels, tanks or pipes, other machine components like gaskets and bushes etc. Natural rubber is resistant to ductile acids, alkalies and salts where it is affected by oils, benzene and other oxidizing media. Synthetic rubber like neoprene, butyl, silicon, vitron etc. are resistant to oils and solvents. Urethrene rubber is used in stirrers, centrifuge baskets, pumps, values, storage vessels, fans etc due to its resistance to acids and alkalies in addition to abrasion. It has good hardness and resilience.

Glass: Glass has high compressive strength but low tensile strength. It is very brittle and can be damaged by thermal shocks. It has an excellent corrosion resistant property. Glass enamel (lining) to the steel or iron vessels gives strength and corrosion resistance as well.

Carbon: When it is impregnated with chemically resistant resins, it can be used in heat exchangers for special purpose. It has a high conductivity and a good resistance to most chemicals, except oxidising acids of >30% concentration. In conventional shell and tube heat exchangers carbon tubes can be successfully used.

Nomenclature

c : cost of the material, Rs/kg

CR : cost rating used for compensation

P : density of the material, kg/m^3

σ_d : design stress, N/mm^2

3

Pressure Vessels

3.1 Introduction

Most of the pressure vessels used as process equipment are either cylindrical or spherical or some composite of these with domed ends. Openings or nozzles are provided with flanged or threaded joints. They are usually supported to elevate from the ground up to the desired level required for the efficient operation. Pressure vessel is not clearly defined but generally it is accepted that any closed vessel over 150 mm diameter subjected to a pressure difference of more than 0·10 N/mm^2 (one bar) should be designed as a pressure vessel. There are two types of pressure vessels *viz.* thin-walled and thick-walled vessels. If the thickness to diameter ratio is less than 0·10, it is a thin-walled pressure vessel and thick-walled if this ratio is higher than this value. The radial stresses in thin-walled vessels are very small and can be neglected in comparison to principal (longitudinal and circumferential) stresses which can be taken as constant over the wall thickness. The magnitude of the radial stress will be significant in thick wall whereas the circumferential stress will vary across the wall. The majority of food processing industrial vessels are thin-walled, the thick-walled being used for high pressures.

The design and fabrication of thin-walled pressure vessels is covered by national standards and codes of practice including their inspection and testing *e.g.*, IS : 2825 is prepared by the Indian Standards Institute. The design should be checked with the standard code irrespective of the design procedure used.

3.2 Operating design conditions

The operating conditions may be classified as the normal conditions resulting from the maximum conditions during operation and transient conditions resulting from operation during starting up or shutting down or the change of loading.

(1) *Design pressure :* During normal operation a vessel is subjected to the maximum pressure called as the design pressure which is about 5 to 10% higher than the normal

working pressure to meet accidental elevated values for the vessels under internal pressure. It is usually adjusted on the relief valve. In case of vessels subjected to external pressure or vacuum, it should be designed for full negative pressure of 0·10 N/mm^2, unless fitted with vacuum relief valve.

(2) *Design temperature* : The fluid contained in the pressure vessel decides the maximum and minimum temperatures. The material of construction of vessel also decides the operating temperature since the material strength is reduced with an increase in temperature. A specified temperature for specified material should be selected for this purpose.

(3) *Material and design stress* : Pressure vessels are constructed from cast iron, low and high alloy steel, aluminium, copper, nickel and their alloys and reinforced plastics. It should be selected compatible under the process environment and suitable for fabrication. Normal design strength or maximum allowable stress should be decided by the operating conditions of temperature and pressure, for the material to be used for construction of the pressure vessels. Different National codes specify safety factors either on the basis of ultimate tensile strength (2.35 to 4.0) and /or on the basis of yield strength (1.5 to 1.7). The allowable stress is to be calculated based on the ultimate tensile strength and also on yield strength and the minimum of the two is to be taken as allowable design stress. Table 3.1 shows typical design stress values for some common materials.

Table 3.1 Allowable stresses for different metals

(A) Carbon and low alloy steel plates

Material specification	*Grade*	*Ultimate tensile stress, N/mm²*	*Allowable stresses (N/mm²) for different temperatures, °C*							
			250	*300*	*350*	*375*	*400*	*425*	*450*	*475*
IS:2002-1975	1	370	95	87	78	75	72	59	43	36
IS:2002-1975	2A	420	98	90	81	77	74	59	43	36
IS:2002-1975	2B	520	121	111	100	95	83	59	43	36
IS:2041-1986	20Mo55	480	143	132	123	119	115	112	108	77
IS:2041-1986	20Mn2	520	140	128	116	110	83	59	43	36
IS:1570-1975	IS Cr90Mo55	500	160	152	144	138	134	130	126	117
IS:1570-1975	C15 Mn 75	420	107	98	89	84	81	59	43	36

(B) High alloy stainless steel plates and tubes (IS : 1570-1961)

Designation	*Allowable stresses (N/mm²) for different temperatures, °C*							
	50	*100*	*150*	*200*	*250*	*300*	*350*	*400*
04 Cr 19 Ni 19	160	142·0	124·0	106·0	99·7	93·5	87·0	80·7
04 Cr 19 Ni 9 Ti 20	160	142·8	125·6	108·3	106·4	106·0	106·0	108·0
04 Cr 19 Ni 9 Nb 40	160	142·8	125·6	108·3	106·4	106·0	106·0	106·0
05 Cr 18 Ni 11 Mo 3	160	145·0	130·0	115·0	112·4	112·3	112·3	111·2
05 Cr 19 Ni 9 Mo 3 Ti 20	160	145·0	130·0	115·0	112·4	112·3	112·3	111·2

(C) Copper and copper alloy tubes

Material specification	*Grade*	*Ultimate tensile stress, N/mm²*	*Allowable stresses (N/mm²) for different temperatures, °C*							
			30	*75*	*100*	*125*	*150*	*175*	*200*	*225*
IS:407-1986	Alloy 1	290	70·3	70·3	70·3	70·0	69·6	57·0	38·3	—
	Alloy 2	290	87·9	86·7	83·0	78·1	72·8	53·8	20·0	—
IS:1545-1984	ISBT 1	—	70·3	70·2	70·3	70·3	69·6	57·0	38·3	—
	ISBT 2	—								
	ISABT	—	84·4	84·4	84·4	84·4	82·8	54·3	25·8	15·8
	ISABZT	—	87·6	86·7	85·3	83·4	80·9	70·9	46·4	31·6
IS:2371-1984	CuNi 21	320	84·4	84·4	84·4	84·4	82·8	54·3	25·8	15·8
	Al2 As									
IS:2501-1983	CuNi 31	420	83·1	80·8	78·9	77·1	75·9	75·8	72·7	71·4
	MNl Fe	—	42·2	41·9	41·3	40·0	34·7	27·1	19·0	—

(4) *Corrosion allowance and wall thickness :* The material of vessel is subjected to the atmosphere and the liquid in operation which may cause corrosion, erosion or scaling of the vessel material. There is no specific rule for provision of corrosion allowance to meet this requirement. However, 2·0 mm allowance is provided for carbon and low-alloy steels where corrosion is not severe and 4·0 mm for the materials in severe conditions. Most of the design codes and standards specify a minimum allowance of 1·0 mm. The minimum wall thickness shall be provided as specified in Table 3.2 depending on the vessel diameter in order to withstand self weight of vessel including any incidental load. Corrosion allowance should be added to the values of thickness mentioned therein Table 3.2.

Table 3.2 Minimum practical vessel wall thickness excluding corrosion allowance

Pressure vessel diameter, mm	*Minimum wall thickness, mm*
1000	3
1000 to 2000	5
2000 to 2500	7
2500 to 3000	8
3000 to 3500	10

(5) *Design loads :* The pressure vessels are generally designed based on elastic analysis. Beyond elastic limit, the material yields and plastic region spreads without the increase in load. This load is termed as collapse load or bursting pressure. Thus, the pressure vessel must be designed to resist gross plastic deformation and collapse under all the conditions of loading. The loads to which the pressure vessels are subjected are classified under two

groups *i.e.*, major loads and subsidiary loads. The major loads are due to design pressure and static head of liquid, self weight of vessel and conténts, wind load, load due to other connections etc. The subsidiary loads may be due to local stresses caused by supports or the connecting pipes, shock loads caused by water hammer or surging of liquid, bending moment due to eccentric loading, differential thermal coefficient of expansion and loads caused by fluctuations in pressure and temperature. However, the vessel may not be subjected to all the loads, but the worst situation must be assessed and vessel should be designed for such condition.

3.3 Thin-walled vessel under internal pressure

The simplest pressure vessel can be designed by parts namely, shell (cylindrical or spherical), heads and closures, nozzle, flanged joint and supports. These are fabricated either from sheets or from welded or seamless pipes individually and then connected by rivets or welding. For radiographically checked welded joints, the efficiency can be taken as 100%, but it is assumed 85% if the checks are carried out at few spots only and from 50 to 85% if the test is not done. For butt riveted joints it is taken as 70 to 85% and lap riveted joints between 50 to 70%.

3.3.1 Cylindrical shell

The three dimensional stresses are produced in the thin wall of a cylindrical shell namely, hoop or circumferential stress, longitudinal (along the axis) stress and the radial stress (normal to the wall). For the vessels with the ratio of wall thickness to the internal diameter is less than 1/20, the axial and hoop stresses (Fig. 3.1) are assumed to be constant over the thin wall, the radial stresses being very small and can be neglected. These tensile stresses can be expressed as below :

hoop stress, $$\sigma_H = \frac{pD_m}{2t} \qquad (3.1)$$

and axial stress, $$\sigma_A = \frac{pD_m}{4t} \qquad (3.2)$$

For the cylindrical shell the minimum wall thickness for given internal pressure can be determined from equation (3.1). Generally, hoop stress is greater of the two principal stresses. The mean diameter is $(D_i + t)$ and from equation (3.1),

$$t = \frac{p(D_i + t)}{2f} \qquad (3.3)$$

Considering the joint efficiency and rearranging equation (3.3),

$$t = \frac{p\,D_i}{2f\,j - p} \qquad (3.4)$$

If the outer diameter of vessel is considered, then,

$$t = \frac{p\,D_o}{2fj + p} \tag{3.5}$$

Standard dimensions for steel plates and nominal outside diameters for process vessels are given in Appendix A and B, respectively.

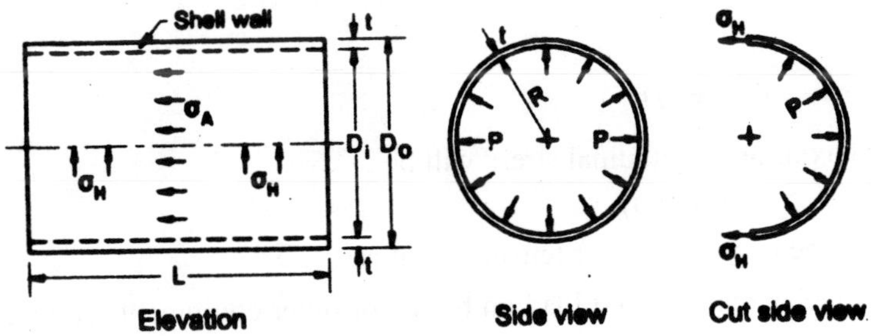

Fig. 3.1 *Stresses in thin walled vessel*

3.3.2 Spherical shell

In case of spherical shell, the hoop stresses and axial stresses are equal and hence, the shell thickness is given by the following equation.

$$t = \frac{p\,D_m}{4fj} = \frac{p\,D_i}{4fj - p} = \frac{p\,D_o}{4fj + p} \tag{3.6}$$

According to IS : 2825-1969, t/D_i should not exceed 0·25 and design pressure should be elevated by minimum 5%.

3.3.3 Vessel subjected to combined bending

Self weight of the vessel and its contents, load due to wind, external loads imposed by offset piping and attached equipment are the other loads to which pressure vessels have to withstand in addition to the internal pressure. The stresses arising from these loads for cylindrical pressure vessels are as below.

(1) Hoop and axial stresses are given by equation (3.1) and (3.2), respectively as,

$$\sigma_H = \frac{p(D_i + t)}{2t} \tag{3.7}$$

and $$\sigma_A = \frac{p\,D_i}{4t} \tag{3.8}$$

(2) In case of vertical cylindrical vessel compressive stress due to self weight of vessel and its content is given by

$$\sigma_W = \frac{W}{\pi t (D_i + t)} \tag{3.9}$$

(3) Bending stresses resulting from wind loads on tall vertical vessels, dead weight and wind loads on other connections to the vertical vessels and self and contents load for horizontal vessels, which may either be tensile or compressive depending on location and given by

$$\sigma_B = \frac{M}{Z} = \frac{M}{\pi t D_i^2} \tag{3.10}$$

Total axial or longitudinal stress will be,

$$\sigma_L = \sigma_A + \sigma_W + \sigma_B \tag{3.11}$$

which may be compressive or tensile in nature.

(4) Mechanical or wind load on piping or other equipments connected to the vessel may cause torsional shear stresses (σ_T) due to the torque (T) produced about the vessel axis which is given by,

$$\sigma_T = \frac{2T}{pt D_i D_m} = \frac{2T}{\pi t D_i (D_i + t)} \tag{3.12}$$

The above stresses can be combined to yield the two dimensional equivalent stress (σ_E) according to the distribution energy theory or shear energy theory, also known as Hencky Von Mises theory. As per this theory, the material failure under the combined stresses will take place when the elastic strain energy required for the material distortion reaches the energy required to yield under simple tension or compression. The combined stresses and the material deformation must be within the elastic limit and hence, the following resulting σ_E should be divided by the factor of safety.

$$\sigma_E = \left[\sigma_H^2 - \sigma_H \cdot \sigma_L + \sigma_L^2 + 3\sigma_T^2\right]^{0.5} \tag{3.13}$$

Some conditions are required to be satisfied for the satisfactory design of pressure vessel *e.g.*, tensile σ_L and σ_E < allowable σ_H and compressive σ_L < allowable σ_C, where, σ_C is the axial compressive stress which does not cause the failure of shell by wrinkling, given by the following equation.

$$\sigma_C = \frac{t E}{6 D_o [3(1 - \nu^2)]^{0.5}} \tag{3.14}$$

The values of Poisson's ratio and modulus of elasticity for vessel wall materials are given in Appendix C and D, respectively. If any of these conditions is violated, the wall thickness should be increased so that the conditions are satisfied. The other allowances for corrosion, supports, temperature or eccentric loading shall be added to the above thickness.

3.3.4 Heads and closures

Various types of heads and closures are employed for closing the cylindrical shells and put in to operation. They are fixed to the shell by means of welding, riveting or bolting. The principal five types are (*a*) flanged, (*b*) flanged and shallow dished, (*c*) flared and dished, (*d*) standard dished or torispherical heads, (*e*) ellipsoidal heads, (*f*) hemispherical heads and (*g*) conical heads.

Flanged only and flanged and shallow dished heads (Fig. 3.2*a* and 3.2*b*) are mainly used for horizontal cylindrical storage vessels at atmospheric pressure. They are also used for manholes or channel covers in heat exchangers. Although fabrication cost is low, the material cost is high for these covers and are uneconomical for a particular diameter and operating conditions. The central portion can be rounded to form a flared and dished head [Fig. 3.2(*c*)]. By turning a small radius at corners of flat plates a formed flat head is manufactured which reduces the local stress to some extent. They are easy to fabricate and have application in storage and low pressure horizontal vessels and bottom heads of vertical vessels resting on firm surface like concrete slab.

Standard dished or torispherical heads [Fig. 3.2 (*d*)] are widely used in pressure range between 0·1 to 1·5 N/mm^2 gauge. For pressure above 1·0 N/mm^2 the cost of these heads should be compared with ellipsoidal heads [Fig. 3.2(*e*)] for selection since the latter are economical to use as compared with the standard dished heads. Ellipsoidal heads are formed on dies usually 2:1 ratio of major to minor axis. Their strength is nearly equivalent to that of the seamless cylindrical shell with the respective inner and outer diameters. Hemispherical head [Fig. 3.2(*f*)] is the strongest head to withstand about twice the pressure of a torispherical head of the same thickness and diameter. However, its manufacturing cost is higher than any other type of head. In this costly head, the material is most efficiently used in case of high pressure vessels.

Conical heads [Fig. 3.2(*g*)] of apex angle about 60° are most commonly used for removal of solids from evaporators, crystallizers, spray dryers, self-emptying grain bins, silos etc. The material is collected on conical bottom and is easily removed from the equipments. In cylindrical pressure vessels of different diameters, this head is used as reducer providing for smooth transmission resulting in a reduced stress concentration at the junction of such shells.

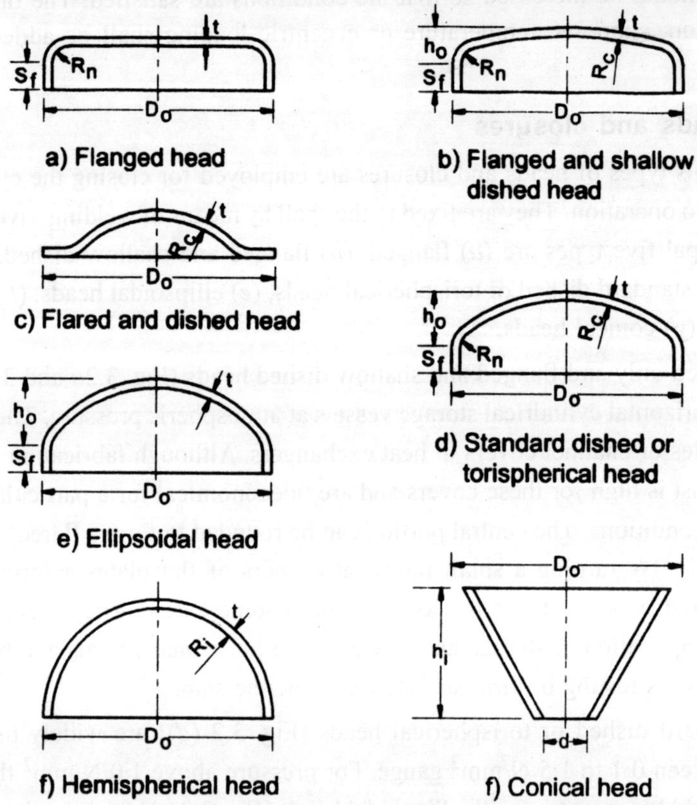

Fig. 3.2 *Types of formed head*

Most of the heads consist of three parts integral with each other *i.e.,* straight flange, corner torus and central dished portion (Fig. 3.3). The straight flange connects head with cylindrical shell. Central dished portion is a part of sphere with a radius of curvature called as crown radius (R_c). This part has the same strength as that of the spherical vessel of the same radius. The two parts are joined by corner torus which converts sphere into a cylinder gradually and has a radius of curvature called as the knuckle radius (R_n). The stress concentration occurs at this junction portion. Larger the knuckle radius, stronger is the torus portion. Generally, knuckle radius is provided larger of the three times head thickness and six per cent of inside head diameter. Minimum length of straight flange should be larger value of three times end thickness or 20 mm, whereas the maximum length may be up to 200 mm in special cases, the standard length being 40 to 100 mm.

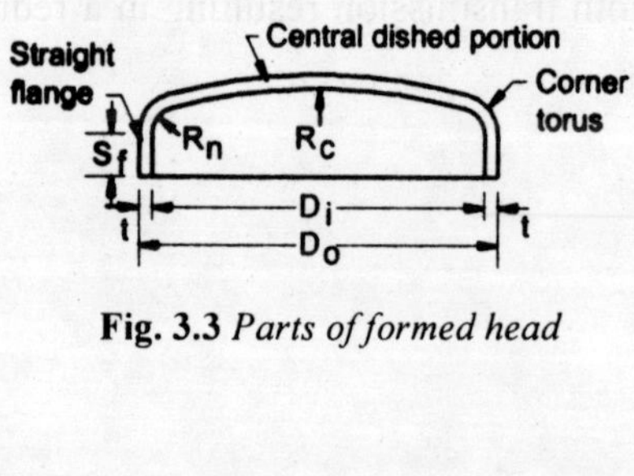

Fig. 3.3 *Parts of formed head*

Design of flat heads

The maximum stress at circumference created in flat plate due to uniform internal pressure and fixed or constraint edges is given by

$$f = \frac{3\,p_i\,R_e^2}{4t^2} = \frac{3\,p_i\,D_e^2}{16t^2} \tag{3.15}$$

and the minimum thickness required under this pressure is given by

$$t_h = C_e\,D_e\left[\frac{p_i}{f}\right]^{0.5} \tag{3.16}$$

The edge fixity constant is assumed from 0·4 to 0·7 depending on the type of fixity of a plate as mentioned below.

(1) Flat plate welded to the shell ($D_e = D_i$)

(*a*) Butt welded, $C_e = 0{\cdot}45$

(*b*) Inside of shell, $C_e \geq 0{\cdot}55$

(*c*) At the end of shell, outside welding only, $C_e = 0{\cdot}7$

(*d*) At the end of shell, inside welding only, $C_e = 0{\cdot}55$

(2) Bolted or riveted end cover with full face gasket D_e = pitch circle diameter (D_p), $C_e = 0{\cdot}42$.

(3) Bolted end cover with a narrow face gasket, $D_e = G$ and C_e is given by,

$$C_e = \left\{0\cdot31 + 190\left[\frac{F_B h_G}{(p_i\,D_e^3)}\right]\right\}^{0.5} \tag{3.17}$$

where, $h_G = 0{\cdot}5\,(D_p - D_e)$.

Design of dished or domed heads

Elliptical or torispherical shapes are most frequently used for cylindrical shells. Due to excessive forming requirement, the hemispherical shape is seldom used. For the determination of wall thickness of formed heads the circumferential stresses are used in the following equations.

(*a*) *Shallow dished and torispherical heads* : The crown radius is equal to or less than the head diameter. The knuckle radius is 0·06 to 0·10 times the outside vessel diameter and thickness is given by,

$$t_h = \frac{p_i\,R_c\,S_i}{2f\,j} \tag{3.18}$$

where, S_i is given by the equation (3.19) below.

$$S_i = 0\cdot 25\left[3 + \left(\frac{R_c}{R_n}\right)^{0\cdot 5}\right] \tag{3.19}$$

(*b*) *Ellipsoidal heads* : The wall thickness of these heads is given by the following equation.

$$t_h = \frac{p_i\, D\, S_e}{2 f\, j} \tag{3.20}$$

where, $S_e = 0{\cdot}25\ (2 + k^2)$ (3.21)

Most common ellipsoidal heads have $k = 2$ and generally does not exceed 2·6, the corresponding values of S_e being 1·5 and 2·19.

(*c*) *Hemispherical heads* : The wall thickness is given by,

$$t_h = \frac{p_i\, D_i}{2 f\, j} \tag{3.22}$$

(*d*) *Conical heads* : The circumferential stress in this case can be computed by using the following expression.

$$f = \frac{p_i\, D_i}{2t \cos \alpha} \tag{3.23}$$

The head thickness can be determined as,

$$t_h = \frac{p_i\, D_i}{2 f\, j \cos \alpha} \tag{3.24}$$

The cone is fixed to the shell by a knuckle with a radius not less than 10% of the internal shell diameter due to occurrence of stress concentration at the junction. Instead of knuckle a compression ring be welded to a shell as an alternate arrangement.

3.4 Thin walled vessel under external pressure

Process vessels subjected to external pressure (vacuum) of less than 0·1 N/mm^2 (1·0 atm) and jacketed vessels fall under this group. This type of vessel may fail by elastic instability or buckling which is the base used for determination of shell wall thickness. The theoretical expression as below will give the critical buckling pressure.

$$p_c = \frac{1}{3}\left[n^2 - 1 + \frac{2n^2 - \nu - 1}{4n^2 Z_L^2 - 1}\right]\cdot\frac{2\,E\,Z_T^3}{(1-\nu^2)} + \frac{2\,E\,Z_T}{(n^2-1)\left(4n^2 Z_L^2 + 1\right)^2} \tag{3.25}$$

where, the Poisson's ratio is usually taken as 0·3 for most of the materials used for shell, $Z_L = L/\pi\, D_o$, and $Z_T = t/D_o$. After dividing p_c in equation (3.25) by a factor of safety, safe external pressure can be obtained. With D_o/L as ordinate and p_e as abscissa, a series of curves can be plotted for $n = 2$ or above. The curve yielding the least value of p_e is then used to find the p_e corresponding to a given value of D_o/L.

It is very cumbersome and time consuming to design shell thickness from equation (3.25). The following simplified correlation is suggested for $\nu = 0{\cdot}3$.

$$p_c = K_c\, E\left(\frac{t}{D_o}\right)^c \tag{3.26}$$

Collapse coefficient is the function of vessel wall thickness, diameter and length (L for vessels without stiffener and L_c for vessels with stiffener) Some values of collapse coefficient and exponent as a function D_o/L are given in Table 3.3.

Table 3.3 Collapse coefficient (K_c) and exponent (c) in equation (3.26)

Collapse D_o/L	Collapse coefficient (K_c)	Exponent (c)
0·1	0·185	2·60
0·2	0·224	2·54
0·3	0·229	2·47
0·4	0·246	2·43
0·6	0·516	2·49
0·8	0·660	2·48
1·0	0·879	2·49
1·5	1·572	2·52
2·0	2·364	2·54
3·0	5·144	2·61
4·0	9·037	2·62
5·0	10·359	2·58

Shell thickness as determined by equation (3.26) must be checked for the plastic deformation for which equation (3.1) equally applies for external pressure with effective length less than critical length resulting in compressive stresses as,

$$\sigma_c = \frac{p_c\, d_o}{2t} \tag{3.27}$$

Substituting allowable stress in equation (3.27), the expression for safe external pressure against the plastic deformation is,

$$P_e = 2f\left(\frac{t}{D_o}\right) \tag{3.28}$$

3.4.1 Ovality factor

The cylindrical shell may not have a perfect circular shape in practice which is detrimental to the vessel strength under external pressure. A deviation from a true circular cross section equal to the shell thickness will reduce the critical buckling pressure by about 50%. This out-of-roundness is measured by an ovality factor (U) given by

$$U = \frac{2(D_{max} - D_{min})}{(D_{max} + D_{min})} \tag{3.29}$$

In case of flat spots or dents on shell surface, the ovality factor is given by,

$$U = \frac{4a}{D_o} \tag{3.30}$$

The value of ovality factor should not exceed 1·5% for vessels under external pressure. For vessels having $L < 0{\cdot}2\ D_o$, the ovality factor is neglected and equation (3.28) should be used for the design. However, in case of vessels with $L > 0{\cdot}2\ D_o$, the following expression should be used.

$$P_e = 2f\left(\frac{t}{D_o}\right)\left[1 + \frac{1\cdot5\,U\left(1 - \dfrac{0\cdot2\,D_o}{L}\right)}{100\left(\dfrac{t}{D_o}\right)}\right]^{-1} \tag{3.31}$$

3.4.2 Stiffner rings

Various rolled steel sections such as flat plate, angles, tee, channels or I-sections are either welded, riveted or bolted to the shell as stiffener rings which offer resistance to collapsing of vessel wall under external pressure. This saves a considerable weight and material. The critical distance between the stiffeners beyond which the stiffening will not be effective is given by the following expression for $\nu = 0{\cdot}3$.

$$L_c = 1\cdot1\,D_o\left(\frac{D_o}{t}\right)^{0\cdot5} \tag{3.32}$$

The stiffeners should be rigid enough to take the load for a distance $0{\cdot}5\ L_c$ and is measured in terms of the moment of inertia. The load per unit length on ring will be given by,

$$F_r = P_e\, L \qquad [3.33\ (a)]$$

and the critical buckling load on ring will be,

$$F_c = \frac{24\, E\, I_r}{\left(D_r^3\right)} \qquad [3.33\ (b)]$$

Usually D_r is equal to D_o. The required ring dimensions can be determined by the combination of equation [3.33 (*a*)] and [3.33 (*b*)]

$$P_e L\ (\text{factor of safety}) \le \frac{24\, E\, I_r}{\left(D_r^3\right)}.$$

3.4.3 Spherical shell

Critical buckling pressure and the corresponding critical stress by using 4.0 factor of safety for spherical shell is given by,

$$P_c = \frac{8\, E\,(t/D)^2}{[3\,(1-\nu^2)]^{0.5}} \qquad (3.34)$$

$$F_c = \frac{2\, E\,(t/D)}{[3\,(1-\nu^2)]^{0.5}} \qquad (3.35)$$

For $\nu = 0{\cdot}3$, the above equations will be reduced to,

$$P_c = 4\cdot 84\, E\left(\frac{t}{D}\right)^2 \qquad (3.36)$$

and

$$F_c = 1\cdot 21\, E\left(\frac{t}{D}\right) \qquad (3.37)$$

A safety factor of 6 is to be assumed for computing the permissible stress. For the determination of allowable external working pressure, select the value of t/D first and then obtain the critical pressure and stress which should satisfy the permissible stress in the material and external pressure on wall.

3.4.4 Design of heads

Torispherical, hemispherical and ellipsoidal heads are designed for external pressure similar to that for the internal pressure except that the internal pressure should be taken

1·67 times the external pressure. In addition, for having the elastic stability assuming a suitable safety factor the thickness is given by,

$$t = 4 \cdot 4\, R_c \left[\frac{3\, P_e(1-\nu^2)}{2E}\right]^{0\cdot 5} \qquad [3.38\ (a)]$$

for ν = 0·3, equation [3.38(*a*)] becomes,

$$t = 5 \cdot 14\, R_c \left[\frac{P_c}{E}\right]^{0\cdot 5} \qquad [3.38\ (b)]$$

The external height of the standard dished head is given by,

$$H_o = R_c - [(R_c - 0 \cdot 5\, D_o)\,(R_c + 0 \cdot 5\, D_o - 2\, R_n)]^{0 \cdot 5} \qquad (3.39)$$

Similarly, the inside depth is also calculated using the equation (3.39) considering internal radii and vessel diameter. Then, for vessels without stiffeners, the effective length is given by the expression,

$$L_e = L + \left(\frac{2H_i}{3}\right) \qquad (3.40)$$

Conical heads with apex angle less than 45° under the external pressure are designed as a cylindrical shell with the same diameter as the large end of the cone and length as the axial height of cone. Heads with apex angle from 45° to 120° are designed as cylindrical shell with diameter and length both equal to the diameter of larger end. Those having apex angle more than 120° are designed as flat plate with diameter of the larger end.

3.5 Compensation for openings in vessel

Process pressure vessels are generally provided with the nozzles or openings for inlet or outlet connections, manways, vents, drains and for instruments such as pressure and temperature gauges. These may be located either on shell or heads depending on the function to be performed. Such openings weaken the wall and give rise to stress concentration at the periphery of the hole. To compensate for this effect, the wall thickness is increased in the region adjacent to the opening. According to the method of forming and attachment, the nozzles are classified in three groups namely, (*a*) welded pad or fabricated, (*b*) inset and (*c*) integral or forged nozzles.

Welding a pad or collar around the opening [Fig. 3.4 (*a*)] is the simplest method for compensation. The pad outer diameter is usually 1·5 to 2 times the hole diameter. These nozzles are satisfactorily used for small openings with low and medium pressures. Although, it is the simplest method, sometimes, high thermal stresses may arise due to the poor thermal conductivity of the pad to shell junction. In case of inset nozzle, the reinforcement required can be provided at a branch with or without the pad by allowing the branch to protrude into the vessel [Fig. 3.4(*b*)]. Since, the protruded nozzle may act as a trap for crude and may result into severe corrosion, this arrangement should be carefully provided. Integral or forged nozzles are most effective but are expensive. They can be successfully provided for large openings under severe conditions. These nozzles are provided while fabrication of the shell or head while casting [Fig. 3.4 (*c*)]. Because of high cost such nozzles are used in special cases only. These can also be formed in a specific shape and size, preferably with flanges.

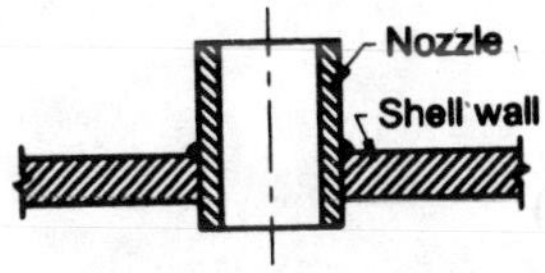

a) Inset nozzle

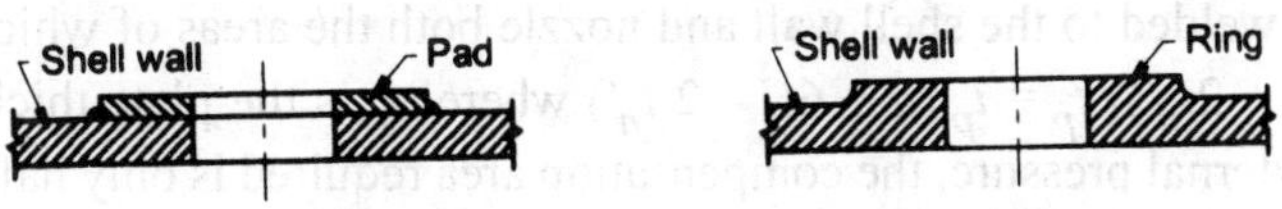

a) Welded pad b) Forging ring

Fig. 3.4 *Types of compensation for openings*

The reinforcement required is determined by three methods, namely, (*a*) area for area method, (*b*) controlled maximum stress method and (*c*) experimental yield method. The area for area method, as suggested by the codes, is the most simple and reliable method for providing the compensation. The other methods are described in codes and should be referred thereon, if required. The effective area of compensation is shown in Fig. 3.5 which has the limits of *abce*. If r is the radius of opening, d is its diameter ($d = 2r$). Let, t_s and t_n be the actual thickness of shell/head and nozzle, respectively; t_s' and t_n' be the theoretical minimum thickness of shell/head and nozzle, respectively calculated with suitable joint efficiency ; h_o and h_i be the height of compensation in branch wall outside and inside shell/head measured from outside and inside surfaces, respectively of compensation ring or vessel wall as the case may be and C be the corrosion allowance.

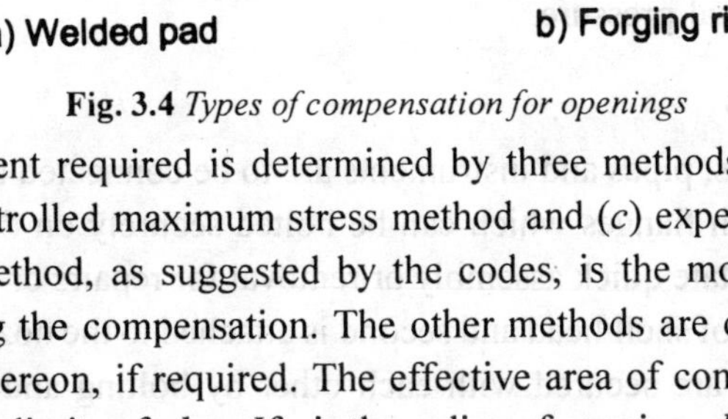

Fig. 3.5 *Boundary limits for reinforcing pad*

The maximum horizontal distance $ab = 2d = 4r$ and the maximum vertical side ae = 6 t_s or 3·5 t_s + 2·5 t_n whichever is less. If the compensation is provided by nozzle only, $h_o = h_i = 2{\cdot}5\ t_n$ and if by nozzle and ring, then $h_o = 2{\cdot}5\ t_s$. Referring to Fig. 3.5 let A be the basic area being reduced due to opening and A' is the area available for compensation, then,

$$A = dt_s$$

and $$A' = A_s + A_o + A_i$$

where,

$$A_s = d\,(t_s - t_s' - C)$$

$$A_o = 2\,h_o\,(t_n - t_n' - C)$$

and $$A_i = 2\,h_i\,(t_n - 2\,C)$$

If $A > A'$, an additional area of $(A - A')$ shall be provided by way of compensation ring (Fig. 3.5) welded to the shell wall and nozzle both the areas of which is $A_p = [2\,(d + 2C) - (d - 2C + 2t_n')]\,t_p = t_p\,(d + 6C - 2\,t_n')$ where, t_p is the plate thickness. In case of vessels with external pressure, the compensation area required is only half of that required for vessels with internal pressure.

3.6 Flanged joints

When the nozzles, pipes and instruments are to be connected to the pressure vessels, they are provided with flanges which can be bolted securely on the shell or head walls. Such provisions facilitate quick assembly or removal for repairs or so. It consists of a pair of flanges, one is part of shell/head and second is attached to the nozzle, pipe or instrument to be mounted. Both are secured with each other by bolting and a gasket between the adjoining faces to make the joint leak-proof.

3.6.1 Types of flanged joints

There are mainly five types of flanges, namely, (*a*) welding neck flanges, (*b*) slip-on flanges, (*c*) lap-joint flanges, (*d*) screwed flanges and (*e*) blind or blank flanges, as shown in Fig. 3.6.

Welding neck flanges [Fig. 3.6(*a*)] normally manufactured by forging, consist of a long tapered hub between the flange ring and the welded joint. Since the hub provides gradual transition reducing the stress concentration, they are the strongest flanges and hence, are preferred for severe service conditions of temperature, pressure, shear and vibrations. Figure 3.6 (*b*) shows the slip-on flanges which are slipped on to the pipe or nozzle and are lap-welded externally or internally by set back of pipe up to 2 mm. These flanges are cheaper, easy to align but poor resistance to shock and vibration and have 1/3 to 2/3 life that of corresponding welding neck flanges. Lap-joint flanges [Fig. 3.6 (*c*)] are used with lap stub for piped work. They can be easily aligned thus, simplifying the erection work of large sized (diameter) vessels. They can also be adopted where frequent dismantling is required either for inspection or cleaning. These flanges can withstand the pressure without leakage. However, they are costly and have only 1/10 of fatigue life as compared with the welded neck flanges. Screwed flanges [Fig. 3.6 (*d*)] are used to connect screwed fittings to flanges and sometimes, for the connection of difficult-to-weld alloy pipes. Blind or blank flanges are flat plates used to blank off the process vessel openings such as inspection ports, hand holes and also ends of pipes and valves. High bending stresses can be absorbed by the blind flanges.

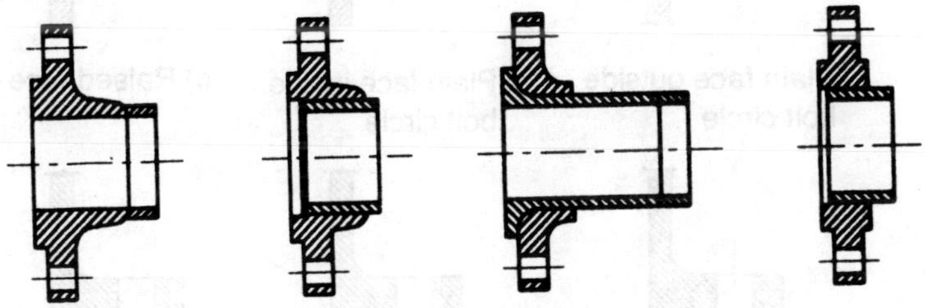

(*a*) Welding neck (*b*) Slip on (*c*) Lap joint (*d*) Screwed

Fig. 3.6 *Types of flanges*

3.6.2 Flange facings

The flange faces are formed to certain finish level and particular shape to accommodate the gasket according to which they are categorized into two groups *i.e.,* wide face flanges where gasket extends the full width of the flange face used for less than 2·0 N/mm^2 pressure and 220 °C temperature and for soft gaskets [Fig. 3.7 (*a*)] and narrow faced flanges where the gasket does not extend beyond the inside of bolt holes [Fig. 3.7 (*b*)]. The later type is most commonly used due to assurance of leak proof joints. These have various types of faces as shown in Fig. 3.7 (*b*) through 3.7 (*f*).

Plain or full flat face flanges [Fig. 3.7 (*b*)] are used for low pressure service and are simple and inexpensive. Both the flanges are identical. Gasket can blow out if pressure exceeds. Raised face flanges [Fig. 3.7 (*c*)] ensure better grip on gasket and no metallic contact even though the gasket is completely pressed. One surface is raised while the other is recessed in case of male and female flanges [Fig. 3.7 (*d*)] which are suitable for high internal pressure and are more expensive than raised face type. Figure 3.7 (*e*) shows a tongue and groove type flange which are most suited for both high internal pressure and vacuum and are expensive. While dismantling the tongue may damage if not handled carefully. Ring type flanges [Fig. 3.7 (*f*)] has groove in both the faces where specially made gaskets are required. This type is less expensive than tongue and groove type. Since the gasket is completely pressed in the circular or trapezoidal cross-sectional groove, there is no fear of blow out of the gasket in internal pressure as well as vacuum.

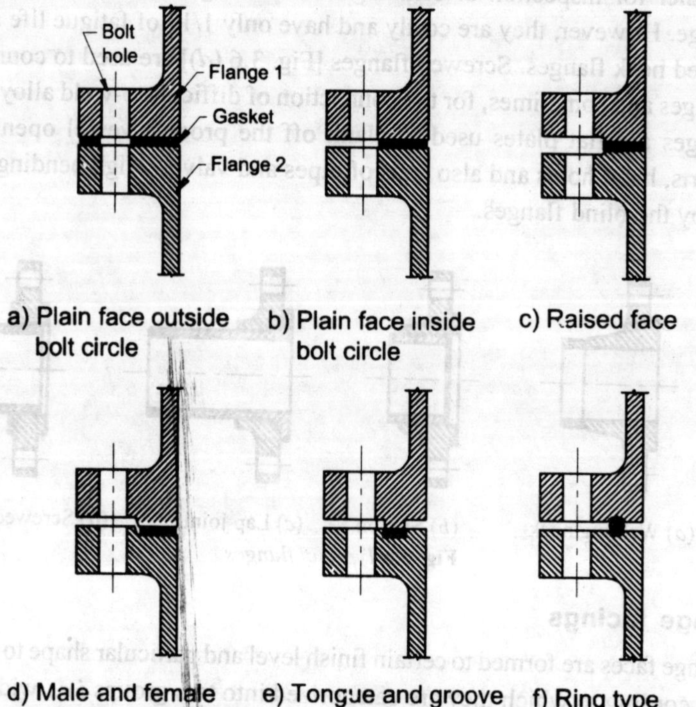

Fig. 3.7 *Flange facing types*

3.6.3 Gaskets

Gaskets are used between two adjacent flange faces and held tight by means of bolts clamping the flanges together. It is very expensive to get fine machined surfaces which can make a leak-proof joint and hence, gaskets made of semi-plastic materials are used. Such materials are deformed and flow under the load to fill the surface irregularities between the flange surfaces, still retaining sufficient elasticity to cope up with the changes in flange alignment occurring under the load. The quantity of force required for tightening the bolts to cause gasket flow is known as yield or seating force expressed as unit stress and is independent of the internal shell pressure. The hydrostatic end force on flange will tend to separate it out. If this force exceeds bolt-load force below a critical value, either leakage may occur or gasket may be blown out. This difference known as the residual gasket force must be less than that required to prevent the leakage. The ratio of gasket stress to the internal pressure is called as the gasket factor, which is the property of the construction material and is independent of internal pressure over a wide range.

For selection of the gasket material, the process conditions (pressure, temperature and corrosion due to process fluid shown in Table 3.4), type of flange and flange face and repetition of dismantling and assembly of flange should be taken into consideration. Gasket material should not form a permanent bond with the flange faces. Table 3.5 gives some common types of gasket materials along with gasket factor and minimum design seating stress and Table 3.6 shows the effective gasket seating width (b) for different flange faces. The effective width is less than the actual gasket width in contact (N). Initially, for preliminary assessment of gasket inner diameters and outer diameter, the following relationship may be used.

$$\frac{G_o}{G_i} = \left[\frac{y - p_i m}{y - p_i\,(m+1)}\right]^{0.5} \tag{3.41}$$

Table 3.4 Commonly used gasket materials under sufficient operating conditions

Pressure, MN/m²	*Temperature, °C*	*Material to be used*
< 1·0	<80	Asbestos or rubber sheet
up to 2	up to 250	Compressed asbestos sheet and various metallic reinforced asbestos sheet
up to 2	up to 400	Corrugated metal-asbestos gaskets, aluminium/copper gaskets
>2	>400	Plain metal gaskets

Table 3.5 Materials for gasket and their design seating stress and gasket factor

Gasket material	*Design seating stress (y), N/mm²*	*Gasket factor, (m)*
1. Rubber with cotton fabric insertion	2·76	1·25
2. Rubber with asbestos fabric insertion, with or without steel wire reinforcement		
(*a*) Single ply	25·50	2·75
(*b*) Double ply	20·00	2·50
(*c*) Three ply	15·25	2·25
3. Vulcanized rubber sheet (hardness >701HR:))	1·38	1·00
4. Asbestos with suitable binder for operating conditions		
(*a*) Thickness 0·8 mm	44·85	3·50
(*b*) Thickness 1.6 mm	25·50	2·75
(*c*) Thickness 3.2 mm	11·00	2·00
5. Corrugated metal, asbestos inserted or asbestos filled corrugated metal jacket		
(*a*) Soft aluminium	20·00	2·50
(*b*) Soft copper or brass	25·00	2·75
(*c*) Stainless steel	45·00	3·50
6. Asbestos filled flat metal jacket		
(*a*) Soft aluminium	38·00	3·25
(*b*) Soft copper or brass	45·00	3·50
(*c*) Stainless steel	62·50	3·75

Note : Provide 10 mm minimum actual gasket width.

Table 3.6 Gasket width for different flange facings

Flange face type	*Basic gasket seating width (b_o)*	*Effective gasket seating width (b)*
1. Plain face	$b_o = 0{\cdot}5$ N	$b = b_o$ for $b_o \leq 6{\cdot}3$ mm
2. Raised face	$b_o = 0{\cdot}5$ N	$b = 2{\cdot}5\,(b_o)^{0{\cdot}5}$ for $b_o > 6{\cdot}3$ mm
3. Male female face	$b_o = 0{\cdot}5$ N	
4. Tongue and groove face	$b_o = N + W_t$	
5. Ring type face	$b_o = 0{\cdot}125\, W_r$	

The product $p_i\, m$ in equation (3.41) is the unit load required to compress the gasket under operating conditions. Normally, for gasket width up to 20 mm, the gasket thickness of 3 mm is used, for 20 to 30 mm gaskets thickness should be 4 mm and above 30 mm, it should be 5 mm. If larger seating stress is desired, thickness less than 3 mm can be used.

3.6.4 Design of bolts

Bolt spacing is the centre-to-centre distance of the two adjacent bolts in a circle called as the pitch circle. Bolt spacing, from practical point, should neither be too large to avoid deflection of flange nor too close to allow the application of a spanner. The following empirical expression can be used to determine the maximum bolt spacing.

$$B_s = \frac{2\, d_b + G\, t_f}{(m + 0\cdot 5)} \tag{3.42}$$

The minimum value of bolt spacing should not be less than $2{\cdot}5\, d_b$ for smaller bolt sizes. Table 3.7 gives the recommended bolt spacing and radial distance from pitch circle diameter to outer edge of hub. Number of bolts should be multiple of four for uniform tightening. The pitch circle diameter should be minimum to avoid flange stresses as well as to accommodate given number of bolts and their spacing as below.

$$D_p = \frac{n_b\, B_s}{\pi} \tag{3.43}$$

For providing the necessary radial clearances the pitch circle diameter should satisfy the following condition.

$$D_p = G_i + 2\,(1{\cdot}415\, t_h + r) \tag{3.44}$$

If both the values from equation (3.43) and (3.44) are equal, the design is optimum. However, larger value should be selected, if both are different. The bolt material may be selected considering the allowable stress and working temperature conditions from Table 3.8. Bolt hole in the flange should be 2-3 mm larger than bolt diameter for adjustment during assembly and the flange outer diameter should be minimum 20 mm larger than the sum of D_p and d_b to accommodate the bolt head.

Table 3.7 Recommended bolt spacings and radial distance from pitch circle diameter to hub outer edge

Bolt size	*Bolt spacing* (B_s), *mm*	*Radial distance* (*r*), *mm*
M 12 × 1·5	30-75	20
M 14 × 1·5	35-75	22
M 16 × 1·5	40-75	25
M 18 × 2	45-75	27
M 20 × 2	50-75	30
M 22 × 2	55-75	33
M 24 × 2	60-75	35
M 27 × 2	68-75	38
M 30 × 2	75	44
M 33 × 2	77	47
M 36 × 3	80	50
M 39 × 3	86	52
M 42 × 3	91	55
M 45 × 3	96	57
M 48 × 3	102	61
M 52 × 3	110	65
M 56 × 4	118	69
M 60 × 4	126	75
M 64 × 4	134	80
M 68 × 4	142	85
M 72 × 4	150	89
M 76 × 4	158	93
M 80 × 4	166	96

In order to determine the size and number of bolts, the bolt load under operating condition and *bolting-up condition* is determined and higher value should be used for computation. Bolt *load* under internal pressure is given by,

Table 3.8 Allowable design stresses for various bolting materials for working temperature

Material	*Allowable design stress for metal temperature (°C) less than, N/mm²*						
	50	*100*	*200*	*250*	*300*	*350*	*400*
(1) Hot rolled carbon steel	57·3	55·1	53·5	47·6	—	—	—
(2) 1% Cr Mo steel	193·0	181·5	167·7	158·9	154·5	148·3	140·0
(3) 5% Cr Mo steel	138·0	138·0	138·0	138·0	138·0	138·0	138·0
(4) 13% Cr Ni steel	176·0	162·0	140·5	134·0	126·5	119·0	104·5
(5) 18·8 Cr Ni steel	129·0	109·0	85·0	78·5	76·0	73·2	72·0
(6) 18 Cr 2 Ni steel	212·0	195·0	170·0	161·0	152·0	144·0	127·0

$$P_o = P_p + P_g = [0{\cdot}25\ \pi\ G^2\ p_i] + [2\ \pi\ G\ b\ m\ p_i] \tag{3.45}$$

The diameter at location of gasket load reaction (G) is taken equal to mean diameter of gasket contact face if $b_o \leq 6{\cdot}3$ mm or equal to gasket outside diameter contact face minus 2 b, if $b_o > 6{\cdot}3$ mm. Therefore, bolt load under internal working pressure is,

$$P_o = \pi\ G\ P_i\ (0{\cdot}25\ G + 2\ b\ m) \tag{3.46}$$

The bolt load under bolting-up condition is given by,

$$P_b = \pi\ G\ b\ y \tag{3.47}$$

Theoretical bolt area required under each condition is obtained by dividing P_o and P_b by the allowable stress at operating temperature (Table 3.8) and at ambient temperature, respectively. The design is ideal if bolt area in both the cases is equal. Bolt area larger of the two should be considered for the design. Area of a bolt is equal to A_d divided by the number of bolts. Considering the pitch circle diameter conditions in equation (3.43) and (3.44) actual bolt area should be provided (A_b) such that $A_b > A_d$. While providing A_b the root area of the bolt should be considered. The following condition must be satisfied to prevent the gasket damage, or otherwise, change the gasket material (Table 3.4) and repeat the procedure.

$$\frac{A_b\ f_a}{\pi G N} < 2y \tag{3.48}$$

3.6.5 Flange thickness

An approximate flange thickness is given by the following expression,

$$t_f = G\left[\frac{P_i}{(K f)}\right]^{0.5} + C \tag{3.49}$$

where, $$K = \left[0{\cdot}3 + \frac{1{\cdot}5\ P_m\ h_G}{P_p G}\right] \tag{3.50}$$

The flange thickness as determined by equation (3.49) should have a minimum value *e.g.*, for carbon steel flanges, the minimum thickness should be between 20 and 40 mm. IS : 2825 gives a rigorous method for the determination of the flange thickness by different stresses.

Example 3.1 : *Determine the standard plate thickness of a process vessel cylindrical wall made of Grade 2B quality steel of IS code which have allowable stress of 120 N/mm² at working temperature with joint efficiency of 0·85 and working internal pressure of 0·476 N/mm², if diameter is 1250 mm and length 2500 mm. What maximum internal pressure a spherical vessel can take safely, if it is made of the same material and diameter?*

Solution. $f = 120$ N/mm^2, adding 5% extra, $P_i = 0{\cdot}476 \times 1{\cdot}05 = 0{\cdot}5$ N/mm^2, D_o = 1250 mm, $j = 0{\cdot}85$

Using equation (3.5) and substituting the above values,

$$t = \frac{0\cdot5 \times 1250}{(2 \times 120 \times 0\cdot85 + 0\cdot5)} = 3\cdot06 \text{ mm}$$

Providing the corrosion allowance of 2·5 mm, the total wall thickness will be, 5·56 mm. Hence, provide 6 mm standard plate for fabrication of the given vessel. Then actual wall thickness $t_a = 6 - 2{\cdot}5 = 3{\cdot}5$ mm and $D_i = 1250 - (2 \times 6) = 1238$ mm. Then, $t/D_i = 3{\cdot}5/1238 = 2{\cdot}81\ (10)^{-3}$ which is much < 0·25 hence, the design is OK.

For spherical vessel rearranging equation (3.6) for the internal pressure,

$$P_i = \frac{4\,f\,j\,t}{(D_o - t)}$$

$$= \frac{4 \times 120 \times 0\cdot85 \times 3\cdot5}{(1250 - 3\cdot5)}$$

$$= 1\cdot14 \text{ N/mm}^2$$

With less 5% working pressure, $P_i = 1{\cdot}14 \times 0{\cdot}95 = 1{\cdot}08$ N/mm^2

The t/D_i ratio is same in this case also, hence, OK.

Example 3.2 : *A process vessel designed in example 3.1 is to be provided with heads at the ends. Assuming necessary data, compare the thickness of different types of heads.*

Solution. (*a*) *Flanged flat plate welded to inside of the shell*

Given, $C_e = 0{\cdot}55$ and $D_e = D_i = 1238$ mm, $P_i = 0{\cdot}50$ N/mm^2, $f = 120$ N/mm^2, then thickness is given by equation (3.16) as

$$t_h = 0\cdot55 \times 1238 \left[\frac{0\cdot5}{120}\right]^{0\cdot5} = 43\cdot95 \text{ mm}$$

Providing 2·5 mm corrosion allowance total wall thickness is, $t_h = 44{\cdot}95 + 2{\cdot}5 = 46{\cdot}45$ mm where, 50 mm standard plate should be used,

(*b*) *Torispherical head*

Let, crown radius, $R_c = 1200$ mm (equal to or less than head diameter), then $R_n = 1200 \times 0{\cdot}06 = 72$ mm and intensification factor is given by equation (3.19).

$$S_i = 0\cdot25 \left[3 + \left(\frac{1200}{72}\right)^{0\cdot5}\right] = 1\cdot77$$

and $$t_h = \frac{0 \cdot 5 \times 1200 \times 1 \cdot 77}{(2 \times 120 \times 0 \cdot 85)} = 5 \cdot 20 \text{ mm}$$

then $t_h = 5{\cdot}20 + 2{\cdot}5 = 7{\cdot}70$ mm

Provide 8 mm standard thickness of wall for torispherical head.

(*c*) *Ellipsoidal head*

Let, Major axis $D = 1200$ mm and the ratio of major to minor axes of ellipsoidal head be $k = 2{\cdot}2$, then value of S_e is given by using equation (3.21)

$$S_e = 0{\cdot}25\ [2 + (2{\cdot}2)^2] = 1{\cdot}71$$

and $$t_h = \frac{0 \cdot 5 \times 1200 \times 1 \cdot 71}{(2 \times 120 \times 0 \cdot 85)} = 5 \cdot 03 \text{ mm}$$

Provide 8 mm standard thickness for this head wall with 2·5 mm corrosion allowance.

(*d*) *Hemispherical head*

From equation (3.22)

$$t_h = \frac{0 \cdot 5 \times 1238}{(2 \times 120 \times 0 \cdot 85)} = 3 \cdot 05 \text{ mm}$$

Provide 6 mm standard thickness for this head wall, providing corrosion allowance.

(*e*) *Conical head*

Let, half apex angle be $\alpha = 30°$, then $\cos \alpha = 0{\cdot}866$ and the thickness can be obtained from equation (3.24)

$$t_h = \frac{0 \cdot 5 \times 1243}{(2 \times 120 \times 0 \cdot 85 \times 0 \cdot 866)} = 3 \cdot 52 \text{ mm}$$

Hence, adding corrosion allowance, provide 6 mm standard thickness for conical head also.

Thus, it is obvious that flat plate of 50 mm is required which is bulky and may be uneconomical from material side as compared with other types of heads. Hemispherical and conical heads require least wall thickness amongst all the heads *i.e.*, 6 mm only for the same condition of internal pressure and vessel diameter. The torispherical and ellipsoidal heads require 8 mm wall thickness.

Example 3.3 : *A fabrication tower made out of Grade 1 plain carbon steel (IS : 2002-1962) operating under vacuum of 0·1 N/mm² at 400 °C has 6000 mm length and 3600 mm diameter containing collection trays at 900 mm spacing. Suggest what standard plate thickness should be used for the fabrication of this vessel without stiffener rings.*

Also determine the section to be provided @ 900 mm spaced stiffeners for the same vessel and the shell thickness too.

Solution. Considering a standard dished head, covered by both the ends of given vessel, $R_c = D_o$ and $R_n = 0{\cdot}1\, D_o$. The inner height of head H_i can be determined by using equation (3.39), considering $D_i = 3600$ mm approximately equal to D_o, $R_c = D = 3600$ mm and $R_n = 0{\cdot}1 \times 3600 = 360$ mm, then,

$$H_i = 3600 - [(3600 - 0{\cdot}5 \times 3600)\,(3600 + 0{\cdot}5 \times 3600 - 2 \times 360)]^{0{\cdot}5}$$

$$= 698 \text{ say, } 700 \text{ mm}$$

and $$L_e = L + \left(\frac{2\,H_i}{3}\right) = 6000 + \left(\frac{2 \times 700}{3}\right) = 6467 \text{ says, } 6470 \text{ mm}$$

and $$\frac{D_o}{L_e} = \frac{3600}{6470} = 0{\cdot}556$$

The shell wall shall be designed for full vacuum of $0{\cdot}1$ N/mm^2, $E = 1{\cdot}69\,(10)^5$ N/mm^2, $k_c = 0{\cdot}457$, $c = 2{\cdot}477$. Using equation (3.26), the shell wall thickness is determined as below.

$$P_e = K_c\, E \left(\frac{t}{D_o}\right)^c$$

or $$0{\cdot}1 = 0{\cdot}457 \times 1{\cdot}69\,(10)^5 \left(\frac{t}{3600}\right)^{2{\cdot}477}$$

then $t = 15{\cdot}11$ mm *i.e.*, provide standard thick plate 16 mm.

This thickness is to be checked for plastic deformation using equation (3.31) where, $f = 72$ N/mm^2 and $U = 1{\cdot}5\%$, then,

$$P_e = 2 \times 72 \left(\frac{16}{3600}\right)\left[1 + \frac{1{\cdot}5 \times 1{\cdot}5\,(1 - 0{\cdot}2 \times 0{\cdot}556)}{100\,(16/3600)}\right]^{-1}$$

$$= 0{\cdot}12 \text{ N/mm}^2$$

Thus, allowable stress in wall is greater than $0{\cdot}1$ N/mm^2 and the computed wall thickness is safe for plastic deformation.

Providing stiffeners at 900 mm spacing, $L = 900$ mm, $D_o/L_c = 3600/900 = 4$, $K_c = 9{\cdot}037$ and $c = 2{\cdot}62$ from Table 3.3. Then using the same equation, thickness is,

$$0 \cdot 1 = 9 \cdot 037 \times 1 \cdot 69\,(10)^5 \left(\frac{t}{3600}\right)^{2 \cdot 62}$$

then $\quad t = 6 \cdot 52$ mm *i.e.* provide 8 mm standard plate thickness.

also $\quad P_e = 2 \times 72 \left(\dfrac{8}{3600}\right)\left[1 + \dfrac{1 \cdot 5 \times 1 \cdot 5\,(1 - 0 \cdot 2 \times 4)}{100\,(8/3600)}\right]^{-1}$

$= 0 \cdot 11$ N/mm^2 *i.e.*, $> 0 \cdot 10$ hence, safe.

For selection of the section for stiffener, moment of inertia is to be determined from conditions arising by combining equation [3.33 (*a*)] and [3.33 (*b*)] as below, using factor of safety as 2·0. Then,

$$I_r = \frac{2\, P_e\, L\, D_r^3}{24\, E}$$

$$= \frac{2 \times 0 \cdot 11 \times 900\,(3600)^3}{24 \times 1 \cdot 69\,(10)^5} = 2277586 \text{ mm}^4 \qquad \text{say, } 228 \text{ cm}^4$$

Provide ISJC 125 channel (I = 270 cm^4) or ISA 110 × 110 × 10 angle (I = 238·4 cm^4) for fabrication of the stiffener rings. However, channel has less weight (7·9 kg/m) than angle (16·5 kg/m) as selected above. Hence, providing ISJC 125 channel for stiffeners will be economical.

Example 3.4 : *A nozzle of 180 mm diameter has wall thickness of 12 mm fitted with its length 8 mm above and below the wall surface of cylindrical shell having 1600 mm diameter and 30 mm wall thickness. The maximum working pressure inside the vessel is 2·4 N/mm^2 whereas, allowable stress in wall material is 92 N/mm^2. Design the compensation pad to be provided for nozzle fitting, assuming joint efficiency and corrosion allowance as 0·85 and 2·5 mm, respectively.*

Solution. Given, d_i = 180 mm, t_n = 12 mm, $h_o = h_i$ = 8 mm, D_o = 1600 mm, t_s = 30 mm, P_i = 2·4 × 1·05 = 2·52 N/mm^2, j = 0·85 and C = 2·5 mm. Using equation (3.5), shell thickness is,

$$t_s' = \frac{P_i\, D_o}{2 f\, j + P}$$

$$= \frac{2 \cdot 52 \times 1600}{2 \times 92 \times 0 \cdot 85 + 2 \cdot 52} = 25 \cdot 37 \text{ mm}$$

and $t_n' = \dfrac{2 \cdot 52 \times 180}{2 \times 92 \times 0 \cdot 85 + 2 \cdot 52} = 2 \cdot 85$ mm

Area for which compensation required as,

$A = d\,t_s = 180 \times 30 = 5400$ mm^2

Area available for compensation is,

$A_s = d\,(t_s - t_s' - C) = 180\,(30 - 25{\cdot}37 - 2{\cdot}5) = 383$ mm^2

$A_o = 2\,h_o\,(t_n - t_n' - C) = 2 \times 8\,(12 - 2{\cdot}85 - 2{\cdot}5) = 106$ mm^2

$A_i = 2\,h_i\,(t_n - 2C) = 2 \times 8\,(12 - 2 \times 2{\cdot}5) = 112$ mm^2

Total compensated area by shell and nozzle is,

$A' = 383 + 106 + 112 = 601$ mm^2

Area to be compensated by pad is,

$A_p = A - A' = 5400 - 601 = 4799$ mm^2

Now $A_p = t_p\,(d + 6C - 2t_n')$

or $4799 = t_p\,(180 + 6 \times 2{\cdot}5 - 2 \times 2{\cdot}85)$

then $t_p = 25{\cdot}35$ mm

Therefore, provide 28 mm thick standard plate for the compensation pad of d_i = 180 mm and $d_o = 2\,(180 + 2 \times 2{\cdot}5) = 370$ mm.

Example 3.5 : *A process vessel working under 0·95 N/mm² internal pressure and 150 °C having 800 mm outside diameter and 16 mm wall thickness is to be provided with a ring type plain faced flange. If the allowable stresses in flange and bolt materials is 92 and 138 N/mm², respectively, design the flange head.*

Solution. $p_i = 0{\cdot}95 \times 1{\cdot}05 = 0{\cdot}9975$ say $1{\cdot}0$ N/mm^2, $T_i = 150$ °C, $D_i = 800$ mm, t_s = 16 mm, $f = 92$ N/mm^2 and $f_b = 138$ N/mm^2.

Selecting a brass gasket filled with asbestos from Table 3.5, $m = 2{\cdot}75$ and y = 25 N/mm^2. Using equation (3.41), gasket width can be determined. Assuming jacket internal diameter 10 mm larger than D_i, $G_i = 810$ mm, then,

$$\frac{G_o}{G_i} = \left[\frac{25 - 1 \times 2 \cdot 75}{25 - 1\,(2 \cdot 75 + 1)}\right]^{0 \cdot 5} = 1 \cdot 023$$

and $G_o = 1{\cdot}023 \times 810 = 829$ mm, say 830 mm

Hence, least gasket width of 10 mm shall be provided so that G_o = 830 mm, then N = 10 mm and $b_o = 10/2 = 5$ mm

Also $b = b_o = 5$ mm (for $b_o < 6{\cdot}3$ mm)

and $G = G_o - 2b = 830 - 2 \times 5 = 820$ mm

Total bolt load will be as below,

$$P_p = 0{\cdot}25\ \pi\ G^2\ P_i = 0{\cdot}25\ \pi\ (820)^2 \times 1{\cdot}0 = 528102 \text{ N}$$

$$P_g = 2\ \pi\ G\ b\ m\ P_i = 2\pi \times 820 \times 5 \times 2{\cdot}75 \times 1{\cdot}0 = 70843 \text{ N}$$

and $P_o = P_p + W_g = 528102 + 70843 = 598945$ N

also $P_b = \pi\ G\ b\ y = \pi \times 820 \times 5 \times 25 = 322013$ N

Then, the design load is larger of the two *i.e.*, $P_m = 598945$ N.

Bolting area required,

$$A_d = \frac{P_m}{f_a} = \frac{598945}{138} = 4340 \text{ mm}^2$$

Using minimum size of M 12 × 1·5 bolts, root area is,

$$A_r = 0{\cdot}25\ \pi\ (12 - 2 \times 1{\cdot}5)^2 = 63{\cdot}62 \text{ mm}^2$$

$$\text{then, number of bolts} = \frac{A_d}{A_r} = \frac{4340}{63{\cdot}62} = 68{\cdot}22$$

Hence, provide 72 bolts to be multiples of 4 for tightening.

Assuming hub thickness as 10 mm, from equation (3.44),

$$D_p = G_i + 2\ (1{\cdot}415\ t_h + r) = 810 + 2\ (1{\cdot}415 \times 10 + 20) = 878 \text{ mm}$$

Providing $B_s = 36$ mm, from equation (3.43),

$$D_p = \frac{n\ B_s}{\pi} = \frac{72 \times 36}{\pi} = 825 \text{ mm}$$

Since, the difference between the two values of D_p is very small, provide the larger of the two *i.e.*, 878 mm. The outer diameter of flange is given by,

$$d_f = D_p + d_b + N = 878 + 12 + 10 = 900 \text{ mm}$$

For prevention of gasket damage, the condition in equation (3.48) should be satisfied.

$$\text{Then, } \frac{(76 \times 63{\cdot}62) \times 138}{(\pi \times 820 \times 10)} = 24{\cdot}54 \text{ N/mm}^2$$

This is less than $2y = 2 \times 25 = 50$ N/mm^2 hence, gasket is safe.

The flange thickness is determined using equation (3.49),

where $h_G = 0{\cdot}5\ (D_p - G) = 0{\cdot}5\ (878 - 820) = 29$ mm

$P_m = 598945$ N and $P_P = 528102$ N then, from equation (3.50),

$$K = \left[0 \cdot 3 + \frac{1 \cdot 5 \times 598945 \times 29}{820 \times 528102}\right]^{-1} = 2 \cdot 7765$$

and $$t_f = 820\left[\frac{1}{(2 \cdot 7765 \times 92)}\right]^{0.5} + C = 51 \cdot 31 + C$$

Hence, provide 56 mm thick standard plate for flange.

Nomenclature

A	: basic area being reduced due to opening in wall, mm^2
A'	: area available for compensation, mm^2
A_b	: actual bolt area required for head, mm^2
A_d	: design area of bolt for head, mm^2
A_i	: compensation area provided by nozzle internal to vessel, mm^2
A_o	: compensation area provided by nozzle out side to vessel, mm^2
A_p	: compensation area provided by plate, mm^2
A_r	: root mean area of each bolt, mm^2
A_s	: compensation area provided by shell or head wall, mm^2
a	: Maximum value of flat spot or dent on shell wall, mm
B_s	: maximum bolt spacing within the pitch circle, mm
b	: effective gasket seating width, mm
b_o	: basic gasket seating width, mm
C	: corrosion allowance, mm
C_e	: edge fixity constant
c	: exponent
D	: major axis of the ellipsoidal shape, mm
D_e	: nominal plate diameter actually under pressure, mm
D_i	: internal diameter of pressure vessel, mm
D_m	: mean diameter of pressure vessel, mm
D_{max}	: maximum diameter of pressure vessel, mm
D_{min}	: minimum diameter of pressure vessel, mm
D_o	: outer diameter of pressure vessel, mm
D_p	: pitch circle diameter, mm
D_r	: inner diameter of stiffener ring, mm

d	: nozzle diameter, mm
d_b	: bolt diameter
d_f	: outer diameter of flange, mm
d_i	: inner diameter of compensation ring, mm
d_o	: outer diameter of compensation ring, mm
E	: modulus of elasticity of material, N/mm^2
F_B	: bolt load in flat head, N
F_c	: critical buckling load on ring, N
F_r	: load per unit length of ring, N/mm
f	: permissible design stress at design temperature, N/mm^2
f_a	: permissible design stress at ambient temperature, N/mm^2
G	: mean diameter of gasket, mm
G_i	: inner diameter of gasket, mm
G_o	: outer diameter of gasket, mm
H_i	: internal height of standard dished head, mm
H_o	: external height of standard dished head, mm
h_G	: radial distance between gasket load reaction and pitch circle, mm
h_i	: internal height of nozzle that provide compensation, mm
h_o	: external height of standard dished head, mm
I_r	: moment of inertia of stiffener ring cross section, mm^4
j	: joint efficiency factor
K	: flange load factor
K_c	: collapse coefficient
k	: ratio of major to minor axes of ellipse
L	: length of vessel shell, mm
L_c	: critical distance between successive stiffener rings, mm
L_e	: effective length of vessel shell without stiffener rings, mm
M	: total bending moment at given plane, Nm
m	: gasket factor
N	: actual gasket width in contact with flange faces, mm
n	: number of lobes formed at buckling
n_b	: number of bolts provided in a flanged joint

P_b : bolt load under bolting up condition, N

P_g : bolt load due to gasket compression under operating condition, N

P_m : maximum (design) bolt load greater of P_b and P_o, N

P_p : bolt load under operating condition due to design pressure, N

p : design pressure, N/mm^2

p_c :critical buckling pressure, N/mm^2

p_e : safe external pressure, N/mm^2

p_i : internal pressure in vessel, N/mm^2

R_c : crown radius of head, mm

R_e : nominal plate radius actually under pressure, mm

R_n : knuckle radius of head, mm

r : radial distance from pitch circle to outer edge of hub or nozzle, mm

S_e : stress intensification factor for ellipsoidal head

S_i : stress intensification factor for shallow dished or torispherical head

T : torque produced by mechanical or wind load on piping or other equipment connected to the vessel wall, Nm

t : shell wall thickness, mm

t_f : flange thickness, mm

t_h : head wall thickness, mm

t_n : provided nozzle wall thickness, mm

t_n' : calculated nozzle wall thickness, mm

t_p : compensation plate thickness, mm

t_s : provided shell wall thickness, mm

t_s' : calculated shell wall thickness, mm

U : ovality factor

W : weight of vessel itself plus its contents, kg

W_r : width of ring gasket, mm

W_t : width of tongue, mm

y : design seating stress, N/mm^2

Z : section modulus of vessel wall, mm^3

Z_L : ratio of critical distance to the circumferential length of vessel ($L_c / \pi D_o$)

Z_T : ratio of wall thickness to the outer diameter of vessel (t / D_o)

α : half apex angle of conical head, deg

ν : Poisson's ratio

σ_A : axial or longitudinal stress in cylindrical vessel wall, N/mm^2

σ_B : bending stress due to wind load on tall vertical vessel, N/mm^2

σ_C : axial compressive stress in vessel wall, N/mm^2

σ_H : hoop or circumferential stress in cylindrical vessel wall, N/mm^2

σ_L : total axial or longitudinal stress, N/mm^2

σ_T : torsional shear stress due to mechanical or wind load on piping or other equipment connected to the vessel wall, N/mm^2

σ_W : compressive stress on vessel wall due to self weight and its content, N/mm^2

4

Storage Vessels

4.1 Introduction

In a production system, the liquids are stored in order to supply sufficient quantity with an operational continuity ensured while conveying them for various processing operations such as blending, separation, refining, mixing etc. Liquid products are also stored for proper conservation and maintenance of their quality to the acceptable level for transportation and marketing. Vertical cylindrical tanks are widely used for liquid storage at atmospheric pressure. Large sized vessels are used from few litres to thousands of litres. These may be either fabricated in the workshops or vessels so large that they must be fabricated on site due to difficulty in transportation. Generally, mild steel is commonly used for these vessels but rarely stainless steel or aluminium is used in special requirements such as in food industries.

The non-volatile liquids *e.g.*, vegetable oil, refined and unrefined (raw) oil, fruit juices etc. are usually stored in cylindrical vessels having diameter up to 60 m and height 30 m. Small sized tanks may be either horizontal cylindrical with slightly dished or flat heads, supported on two or more saddles ; or rectangular made of flat plates. Since, the flat plates are structurally weak, they can not withstand liquid pressure and need stiffening the walls. Vertical cylindrical tank consists of a vertical shell with flat bottom and plates butt welded. According to the plate thickness, the horizontal joints may be square butt, single or double bevel type (Fig. 4.1). Vertical joints may be square butt, single V or U-butt type. A separate roof is provided with top open so that the tank is free to breath under the changes in temperature and pressure.

4.2 Cylindrical tank design

A cylindrical shell is subjected to the internal loading due to stored liquid with tank assumed full with liquid and external loading due to wind pressure and loads from platforms and elevated walkways. The maximum hydrostatic pressure is exerted at bottom and minimum at top. The shell is constructed by welding plates of appropriate thickness depending on hydrostatic pressure on respective plate for large vessels. For small vessels, plates of uniform thickness can be used for whole height. It is assumed that the bottom plate supports the lowest plate course and each successive are restrained by the respective lower course plates. However, considering the total liquid height in vessel, the design is on safer side. Hydrostatic pressure is calculated by the following expression.

$$p = W_L (H_L - 0{\cdot}3) \qquad (4.1)$$

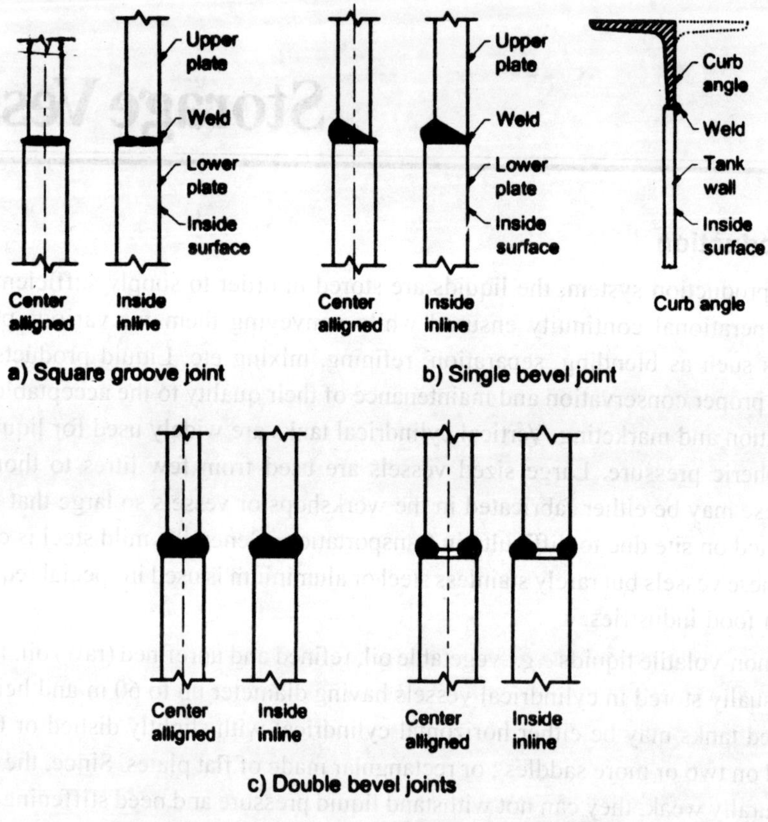

Fig. 4.1 *Horizontal double welded butt joints for shell plates*

The minimum wall thickness required to resist the given hydrostatic pressure can be determined from the following expression.

$$t_c = \frac{p D_i}{2 f j} + C \qquad (4.2)$$

The maximum plate thickness should be 40 mm. Without corrosion allowance, the minimum wall thickness should be at least 5 mm for vessels with < 15 m diameter, 6 mm for 15 to 36 m, 8 mm for 36 to 60 m and 10 mm for > 60 m diameter vessels.

The flat bottom of the tank is fabricated using the rectangular plates cut to form a circle, called as the sketch plates [Fig. 4.2 (*a*)]. An *annular ring* [Fig. 4.2 (*b*)] is to be provided for the tanks with diameter >12 m. All the joints should be lap-welded, with the overlap of not less than five times the thinner plate thickness not exceeding 25 mm. When the entire bottom plate is supported on hard platform, provide 6 mm thick plates including sketch plates or 8 mm thick segmental plates ; since, they are not subjected to the bending stresses. When the tank requires ground clearance, it is supported by angles or beams, the bottom plate is subjected to bending moment due to storage liquid in the tank. The unsupported part of the bottom can be considered as a rectangular plate uniformly loaded. Minimum thickness of bottom plate is given by.

$$t_b = \left[\frac{3wHL_s^2}{4f_i}\right] \tag{4.3}$$

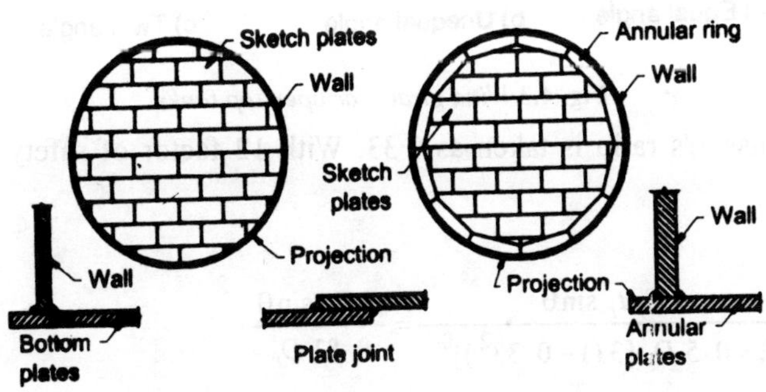

Fig. 4.2 *Details for tank bottom*

The open top tanks require a stiffener near the top course, particularly, outside of the tank shell in order to maintain the shell roundness, called as wind girder. The section modulus of such member is given by,

$$Z = 0{\cdot}059\, D_i^2\, H\, (10)^6 \tag{4.4}$$

Such stiffeners are to be provided from structural sections, formed plate sections or sections built-up by welding, the attachment with shell being shown in Fig. 4.3. The open-top tanks also require a roof which may be conical, dome or umbrella type supported either directly by the tank periphery or on independent columns. Conical roofs are widely used with 1 in 5 or 6 slope or $\tan\theta = 0{\cdot}20$ or $0{\cdot}167$, respectively. A superimposed load of at least $1{\cdot}25$ kN/m^2 is considered on conical roof of storage tanks without any internal pressure, in addition to the other structural load for wind, live loads or snow loads. The stress in cone roof plates is given by,

$$f_r = \frac{p_r\, D_r}{2\, t_r\, \sin\theta} \tag{4.5}$$

In case of larger cones the plates may fail due to wrinkling or elastic instability where, the compressive stress is given by,

$$f_c = \frac{E_r\, t_r}{R[3(1-\nu^2)]^{0.5}} \tag{4.6}$$

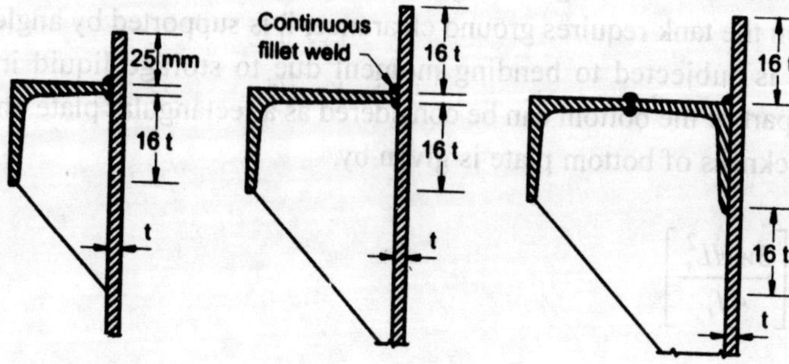

Fig. 4.3 *Wind girder for open top tanks*

Usually Poisson's ratio is taken as 0·33. With 12 factor of safety the permissible stress is,

$$f_a = \frac{E_r\, t_r \sin\theta}{12 \times 0\cdot 5\, D_r\, [3\,(1-0\cdot 33^2)]^{0\cdot 5}} = \frac{E_r\, t_r \sin\theta}{9\cdot 81\, D_r} \tag{4.7}$$

or

$$t_r = \frac{2\cdot 21\, D_r \left(\dfrac{p_r}{E_r}\right)^{0\cdot 5}}{(\sin\theta)} \tag{4.8}$$

The thickness as determined from equation (4.8) should be between 5 and 12 mm welded by continuous fillet weld on top side only supported on purlins spaced < 1·6 m centres. Bracings are also provided in addition to the purlin frame.

Curb angle

For closed top tanks, a top curb angle is provided at the upper edge of shell by a continuous double welded square butt joint (Fig. 4.4). It should have a sufficient cross-sectional area to withstand the load from self-supported roof over the tank top given by

$$A_c = \frac{W_r \cot\theta}{(2\pi f_a)} \tag{4.9}$$

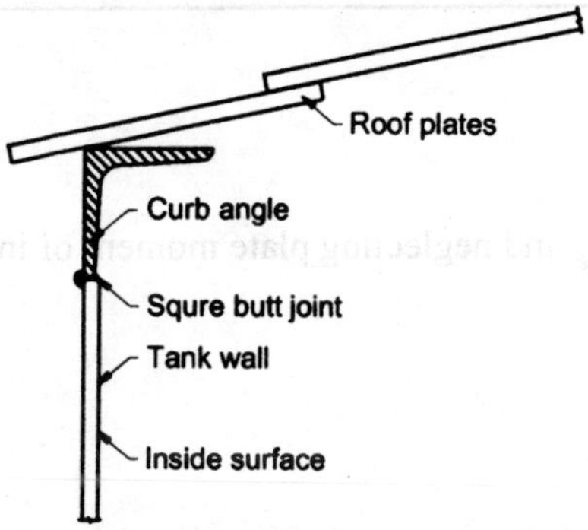

Fig. 4.4 *Curb angle details for roofed tanks*

4.3 Rectangular tank design

For low capacity and negligible pressure of liquids these tanks are preferred since, they are easy to fabricate. The wall thickness is uniform for complete height. The design is based on the use of allowable deflection of plate to within 0·5 times the plate thickness. The rigidity can be achieved by providing stiffeners at the top edge and walls of the tank. The wall thickness without stiffeners for rectangular tank is calculated from the following equation.

$$t_{re} = L\left[\frac{\beta w_L H_L}{f_a}\right]^{0.5} \tag{4.10}$$

and the maximum wall deflection is given by,

$$\delta_{max} = \frac{\alpha\, \omega_L H_L L^4}{E t_{re}^3} < 0.5\, t_{re} \tag{4.11}$$

Values of β and α depending on *H*/*L* ratio as given below.

H/*L* ratio	:	0·50	0·67	1·00	1·50	2·00	2·50	3·00
β	:	0·11	0·16	0·20	0·28	0·32	0·35	0·36
α	:	0·026	0·033	0·04	0·05	0·058	0·064	0·067

When top edge stiffeners are provided, the tank wall plate should be considered as a beam supported at top and bottom with the load increasing from top to bottom. The load on the bottom edge per unit width is given by,

$$F_b = \frac{w_L H_L m}{6} \tag{4.12}$$

The maximum deflection of beam with this loading will be,

$$\delta_{max} = \frac{F_b L^4}{384 EI_s} \tag{4.13}$$

Limiting $\delta_{max} < 0{\cdot}5\ t_{re}$ and neglecting plate moment of inertia, that of the stiffener at edge top is given by,

$$I_s = \frac{F_b L^4}{192 Et_{re}} \tag{4.14}$$

Plate thickness and deflection can be calculated using equation (4.10) and equation (4.11) with replacement of β and α by β′ and α′, respectively and their values are given below.

H/L ratio	:	0·25	0·28	0·33	0.40	0·50	0·67	1·0	1·5	2·0
$\beta'\ (10)^{-2}$	:	2·40	3·10	4·10	5·60	8·00	11·60	16·0	26·00	34·0
$\alpha'\ (10)^{-3}$	:	0·27	0·46	0·83	1·60	3·50	8·30	22·0	43·00	60·0

A large stiffener can be provided horizontally around on four walls of the tank located at 0·577 H from the top of the tank. Also, vertical stiffeners will give higher H/L ratios and reduced thickness for smaller values of L being divided into parts from equation (4.10) and (4.11). The bottom plate of 6 mm thickness should be provided when the entire bottom is supported on hard surface. If the bottom is supported by angles or beams, determine the plate thickness as before from equation (4.3).

The nozzles and mountings to be provided on the storage vessels are (*a*) manholes (*b*) inlet, outlet, over flow and drain connection (*c*) level gauge connections (*d*) combined water draw-off and clean-out sump and (*e*) breather vents. Openings in side wall of tank larger than 60 mm diameter should be reinforced as shown in Fig. 4.5. Breather sizes should be sufficient enough to prevent a pressure or vacuum to built-up beyond the specified range in the tank. It should be 35 mm for 4000-8000 lit capacity tanks, 50 mm for 8000-40,000 lit, 75 mm for 40,000-200,000 lit capacity tanks.

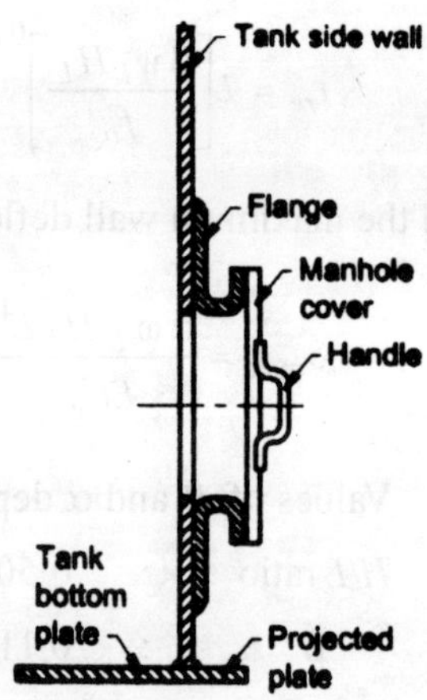

Fig. 4.5 *Manhole reinforcement on side wall of tank*

Example 4.1 : *A circular open-top tank of 988 t capacity is required for storage of raw soybean oil in refinery plant made of carbon steel for which the permissible stress is 124 MN/m² and oil density as 9·8 (10)⁻³ MN/m³. The tank is to be supported on suitable girders placed at 0·75 m centres. The tank is to be provided with a conical roof at top with 1·25 kN/m² superimposed load. The modulus of elasticity and density of the roof plates* *may be assumed as 2·46 (10)^5 MN/m^2 and 62·5 kN/m^3, respectively. Design the tank, if the tank height is restricted to 11·0 m.*

Solution. Capacity = 988 *t*, $w_L = 9{\cdot}8\,(10)^{-3}$ MN/m^3, $H_L = 11{\cdot}0$ m, assuming $j = 0{\cdot}85$, and volume of the tank is,

$$V = \frac{W_L}{w_L} = \frac{988\,(10)^{-3} \times 9{\cdot}81}{[9{\cdot}8\,(10)^{-3}]} = 989 \text{ m}^3$$

and $$D_i = \left[\frac{989 \times 4}{(11\pi)}\right]^{0.5} = 10{\cdot}7 \text{ m}$$

Maximum hydrostatic shell wall thickness will be near the bottom of the tank under hydrostatic pressure given by equation (4.2)

$$p = 9{\cdot}8(10)^{-3}\,(11 - 0{\cdot}3) = 0{\cdot}105 \text{ MN/m}^2 = 0{\cdot}105 \text{ N/mm}^2$$

$$t_s = \frac{0{\cdot}105 \times 10{\cdot}7\,(10)^3}{2 \times 124 \times 0{\cdot}85} = 5{\cdot}33 \text{ mm}$$

With 2 mm corrosion allowance total thickness will be 7·33 mm, hence, provide 8 mm thick plate. Selecting a standard plate size 5600 mm × 1250 mm, the number of courses in tank height will be, 11/1·25 = 8·8 say, 9 courses then total height is 9 × 1·25 = 11·25 m. This extra height of 0·25 m will be used for overlap at joints (0·10 m) and as a free board (0·15 m).

In order to reduce the plate thickness in upper portion of the tank, thickness available is 6 – 2 = 4 mm for height,

$$(H - 0{\cdot}3) = \frac{2f\,jt_s}{(w_L\,D_i)}$$

$$= \frac{2 \times 124 \times 0{\cdot}85 \times 4\,(10)^{-3}}{[9{\cdot}8\,(10)^{-3} \times 10{\cdot}7]}$$

or $$H = 8{\cdot}34 \text{ m}$$

Therefore, provide top 6 courses (6 × 1·25 = 7·5 m height) with 6 mm plate and bottom 3 courses (3 × 1·25 = 3·75 m height) with 8 mm thick plate. Using 6 plates in each course and 2 mm welding allowance at each joint, total circumference is equal to 10·7 π – 6 × 2 (10)$^{-3}$ = 33·60 m, also the circumference is equal to 5·6 × 6 = 33·6 m, OK.

A wind girder is to be provided at the top edge of the tank for which section modulus is,

$$Z = 0{\cdot}059\,D_i^2\,H = 0{\cdot}059\,(10{\cdot}7)^2 \times 11 = 74{\cdot}3 \text{ cm}^3$$

Therefore, provide 2L ISA 110 × 110 × 15 mm as shown in Fig. 4.3.

Bottom plates with 25 mm lap fillet weld joint in staggered manner (Fig. 4.2) and unsupported length of 0·75 m centres, the thickness using equation (4.3) will be,

$$t_b = \left[\frac{3 \times 9 \cdot 8\,(10)^{-3} \times 11\,(0 \cdot 75)^2}{4 \times 124 \times 0 \cdot 85}\right]^{0.5} = 0 \cdot 021 \text{ m} = 21 \text{ mm}$$

Hence, provide 25 mm thick bottom plate with 2·5 mm corrosion allowance.

A conical roof of 11·5 m base diameter shall be provided so that the rain water will spill out away from tank periphery, E_r = 2·46 (10)^5 MN/m^2, w_r = 62·5 kN/m^3 and D_r = 11·5 m. Using maximum plate thickness of 12 mm, total load on roof will be,

$$P_r = 1{\cdot}25 + 12\,(10)^{-3} \times 62{\cdot}5 = 2{\cdot}00 \text{ kN/m}^2$$

Rewriting equation (4.8),

$$\sin\theta = \left(\frac{2 \cdot 21\, D_r}{t_r}\right)\left(\frac{p_r}{E_r}\right)^{0.5}$$

$$= \left[\frac{2 \cdot 21 \times 11 \cdot 5}{12\,(10)^{-3}}\right] \cdot \left[\frac{2 \cdot 00\,(10)^{-3}}{2 \cdot 46\,(10)^5}\right]^{0.5}$$

$$= 0{\cdot}191$$

Then $\theta = \sin^{-1}(0{\cdot}191) = 11{\cdot}01°$

and $\tan\theta = 0{\cdot}194$

which is between 0·167 and 0·2, hence, OK.

Example 4.2 : *A 15·9 m diameter and 12 m height tank is provided for insecticide product storage. Design the tank as closed top assuming specific gravity of liquid as 1·0, allowable stress in wall material as 138 MN/m^2 and tank situated on reinforced cement concrete platform. Provide a fixed conical roof to the tank.*

Solution. D_i = 15·9 m, H = 12 m, w_L = 9·81 (10)$^{-3}$ N/m^3, f = 138 MN/m^2, assume j = 0·85, then the maximum pressure of liquid on wall is,

$$p = 9{\cdot}81\,(10)^{-3}\,(12 - 0{\cdot}3) = 0{\cdot}115 \text{ MN/m}^2 \text{ or } 0{\cdot}115 \text{ N/mm}^2$$

and the plate thickness is,

$$t_s = \frac{0 \cdot 115 \times 15 \cdot 9 (10)^3}{(2 \times 138 \times 0 \cdot 85)} = 7 \cdot 79\,(10)^{-3} \text{ m} = 7 \cdot 79 \text{ mm}$$

Therefore, provide 10 mm thick plate with 2 mm corrosion allowance. Using the plate size 5000 mm × 1250 mm the circumference will be 15·9 π = 49·95 m. Provide 5 plates in a circumference, the extra 0·05 m plate will be used for joint overlap. For reduction of shell wall plate thickness to 8 mm, the upper top height is,

$$(H - 0{\cdot}3) = \frac{2 \times 138 \times 0{\cdot}85 \times 6\,(10)^{-3}}{[9{\cdot}81\,(10)^{-3} \times 15{\cdot}9]} = 9{\cdot}02$$

or $H = 9{\cdot}32$ m

Hence, provide 3 courses of 10 mm thickness, $H = 3{\cdot}75$ m (lower portion of shell) and 12 – 3·75 = 8·25 m *i.e.*, 7 courses of 8 mm plate in upper portion of tank shell. Theoretically, even 6 mm thick plate can also be used. But without corrosion allowance, the minimum wall thickness should be at least 6 mm for vessels with diameter between 15 cm and 36 m. In other words with corrosion allowance the minimum thickness for this vessel is 8 mm. The bottom plate shall be extended 55 mm beyond the shell wall, hence, outer diameter is,

$$D_b = 15{\cdot}9 + 2 \times 10\,(10)^{-3} + 2 \times 55\,(10)^{-3} = 16{\cdot}03 \text{ m}$$

Provide 6 mm thick plates for bottom with 8 mm segmental annular plates [Fig. 4.2 (*b*)] lap welded continuously with 5 times plate thickness *i.e.*, 5 × 6 = 30 mm and for annular plates 55 mm with staggered joints. The minimum width of annular plate will be (2 × 55 + 10 + 380) = 500 mm. The bottom circumference will be 16·03 π = 50·36 m. Selecting 5600 mm × 1000 mm sized plate, 9 plates may be cut from 5 full size plates.

Using the same data from previous example and 12 mm thick plate, with 16·7 m cone base diameter, $P_r = 2{\cdot}00\,(10)^{-3}$ MN/m^2.

then $$\sin\theta = \frac{2{\cdot}21 \times 16{\cdot}7}{12\,(10)^{-3}}\left[\frac{2{\cdot}00\,(10)^{-3}}{2{\cdot}46\,(10)^{5}}\right] = 0{\cdot}277$$

and tan θ = 0·289 which is higher. Hence, it is not advisable to provide self-supporting roof since plate thickness exceeds. Therefore, roof will be supported by columns placed within the tank and purlins of suitable size with 20 mm plate.

Then, $P_r = 0{\cdot}50\,(10)^{-3}$ MN/m^2 and sin θ = 0·186 and tan θ = 0·189 *i.e.*, the slope will be 1:5·28 which is alright.

Example 4.3 : *Design a rectangular tank for oil extraction plant to store about 5000 lit oil having density of 9·72 (10)$^{-3}$ MN/m^3 assuming the permissible stress in wall material as 167 MN/m^2. Also determine the wall plate thickness and size of stiffener if it is provided at the top.*

Solution. Capacity of tank = 5000 lit = 5 m^3. Assuming $H = 2{\cdot}25$ m, the area of a tank will be,

$$A_t = \frac{5}{2{\cdot}25} = 2{\cdot}22 \text{ m}^2$$

Providing tank size of 1·80 m × 1·25 m, the area in plan is 2·25 m^2

$$w_L = 9{\cdot}72\,(10)^{-3}\ \text{MN/m}^3,\ f = 167\ \text{MN/m}^2 \text{ and } L = 1{\cdot}80\ \text{m}$$

then $\dfrac{H}{L} = \dfrac{2{\cdot}25}{1{\cdot}80} = 1{\cdot}25$ for which $\beta = 0{\cdot}24$ and $\alpha = 0{\cdot}05$

From equation (4.10),

$$t_{re} = 1{\cdot}80\left[\frac{0{\cdot}24 \times 9{\cdot}72\,(10)^{-3} \times 2{\cdot}25}{167}\right]^{0{\cdot}5} = 0{\cdot}01\ \text{m} = 10\ \text{mm}$$

Providing 2 mm allowance for corrosion, 12 mm thick plate will be required for walls. For this, the maximum deflection will be,

$$\delta = \frac{0{\cdot}05 \times 9{\cdot}72\,(10)^{-3} \times 2{\cdot}25 \times (1{\cdot}80)^4}{2{\cdot}16\,(10)^5 \times [10\,(10)^{-3}]^3} = 0{\cdot}053\ \text{m}$$

This is higher than 5 mm (0·5 t_{re}), hence provide 22 mm thick plate which gives δ_{max} = 0·49 mm which is safe. Provide 6 mm thick bottom plate assuming the tank placed on hard surface. Providing the stiffener at top edge,

$$F_b = \frac{w_L\, H_L\, m}{6}$$

$$= \frac{9{\cdot}72\,(10)^{-3} \times 2{\cdot}25 \times 1{\cdot}8}{6} = 6{\cdot}56\,(10)^{-3}\ \text{MN/m}$$

for $\dfrac{H}{L} = 1{\cdot}25$, $\beta' = 0{\cdot}21$ and $\alpha' = 0{\cdot}0325$

$$t_{re} = 1{\cdot}80\left[\frac{0{\cdot}21 \times 9{\cdot}72\,(10)^{-3} \times 2{\cdot}25}{167}\right]^{0{\cdot}5}$$

$$= 9{\cdot}44\,(10)^{-3}\ \text{m} \quad \text{say, 10 mm}$$

$$\delta_{max} = \frac{0{\cdot}0325 \times 9{\cdot}72\,(10)^{-3} \times 2{\cdot}25 \times (1{\cdot}8)^4}{2{\cdot}16\,(10)^5 \times [10\,(10)^{-3}]^3} = 0{\cdot}034\ \text{m}$$

Moment of inertia of stiffener is given by equation (4.14)

$$I_s = \frac{F_b\, L^4}{192\, E\, t_{re}} = \frac{6{\cdot}56\,(10)^{-3} \times (1{\cdot}80)^4}{192 \times 2{\cdot}16\,(10)^5 \times 10\,(10)^{-3}}$$

$$= 16{\cdot}60\,(10)^{-8}\ \text{m}^4 = 16{\cdot}60\ \text{cm}^4$$

Therefore, provide 1L ISA 50 × 50 × 6 mm (I = 12·9 cm^4) as a stiffener.

Nomenclature

A_c : cross section area of curb angle, mm^2
A_t : cross section area of tank, m^2
C : corrosion allowance, mm
D_i : internal diameter of storage tank, m
D_r : cone roof base diameter, m
E_r : modulus of elasticity of material, MN/m^2
F_b : load on bottom edge of storage tank, N
f_a : allowable stress in roof material, MN/m^2
f_c : compressive stress in larger cone roof, MN/m^2
f_r : stress in cone roof material, MN/m^2
H : height of tank under consideration, m
H_L : height of liquid in tank, m
I_s : moment of inertia of stiffener ring cross section, mm^4
j : joint efficiency factor
L : length of rectangular tank, m
L_s : unsupported length of bottom plate between two successive supports, m
m : unit width of bottom plate, m
p : hydraulic pressure on tank wall, MN/m^2
p_r : pressure due to roof loading, MN/m^2
R : cone roof base radius, m
t_b : minimum thickness of tank bottom plate, mm
t_c : shell wall thickness of circular tank, mm
t_r : thickness of roof covering material, mm
t_{re} : wall thickness of rectangular tank, mm
W_r : total load on roof, MN
W_L : total weight of liquid in storage tank, t
w_L : density of liquid to be stored in storage tank, MN/m^3
w_r : density of roof material, MN/m^3
Z : section modulus of stiffener, cm^3
α : constant for wall thickness without stiffener
α' : constant for wall thickness with stiffener
β : constant for wall deflection without stiffener
β' : constant for wall deflection with stiffener
θ : cone roof slope angle, deg
δ : maximum deflection of rectangular tank wall, mm
v : Poisson's ratio

5 Heat Exchangers

5.1 Introduction

Heat exchanger is an equipment in which heat is transferred from hot to cold fluid through a separating wall between them. Examples are boilers, condensers, evaporators, automobile radiators, cooling towers etc. The heat transfer may be by direct method without mixing the two fluids. They are separated by metallic or non-metallic surface through which heat is transferred, in the direction perpendicular to the direction of flow.

5.2 Classification

Heat exchangers are mainly divided into three groups according to their working features, as (*a*) regenerators (*b*) recuperators and (*c*) mixed type. The hot and cold fluids flow over the same passage or surface in regenerators. The heat from hot fluid is accumulated on the exchanger surface and is then transferred to cold fluid when it passes over the surface next. The surface temperature varies as it is heated or cooled according to the flowing fluid over it which is also a function of time lag and the flow being transient, the temperature distribution is complex due to which precise calculations are not possible. No mathematical theory has been developed for heat transfer in regenerators. They are mainly used where high temperature waste gas is to be used for air heating *e.g.*, open hearth furnace, blast furnace etc.

Most commonly used heat exchangers are recuperators in which the fluids are separated by a wall and they flow at the same time through the exchanger. The process of heat transfer is through convection from hot fluid to wall, conduction through wall and convection from wall to the cold fluid. Radiation may also occur, if temperature difference is high. An evaporator of ice plant, automobile radiators, boiler preheater, milk chiller of pasteurising plant etc. belong to the recuperator group. The more common types are tubular exchangers and plate exchangers, the widely used being shell and tube exchangers which are described latter. They are further classified according to the temperature distribution or direction of flow of the two fluids into four groups, namely : (*a*) parallel flow exchangers, (*b*) counter flow exchangers, (*c*) cross-flow exchangers and (*d*) condensers or evaporators.

The hot and cold fluids flow in the same direction in parallel flow exchangers (Fig. 5.1) and the temperature difference decreases gradually from inlet to exit, *e.g.*, oil heaters and coolers, water heaters etc. When the two fluids flow in opposite directions (Fig. 5.2), it is called a counter flow exchanger in which the temperature difference between hot and cold fluids remains nearly constant. These exchangers give the maximum heat transfer rates for a given surface area and hence, they are the most suitable devices for heating or cooling the fluids. They may be single or multi-pass depending on the number of passes of fluids once or more than once, respectively. Figure 5.3 (*a*) shows a 1-2 exchanger which has one shell pass and two tube passes and 2-4 exchanger is shown in Fig. 5.3 (*b*). In case of a cross flow exchanger, the two fluids flow at right angle to each other. The arrangements are made such that either each of the fluids is unmixed or one fluid is mixed and the other is unmixed while flowing through the exchanger. The respective examples are automobile radiator [Fig. 5.4 (*a*)] and tubular air heater [Fig. 5.4 (*b*)]. In the former, the air and water flow is unmixed through the tubes but in latter, one fluid (gas or water to be heated) flows through tubes unmixed and air flowing over the tubes is mixed completely. In a condenser, the condensing fluid remains at constant temperature throughout the exchanger whereas the temperature of cold fluid gradually increases as it flow from inlet to exit [Fig. 5.5 (*a*)]. On the other hand, the cold (boiling) fluid remains at the constant temperature in evaporators and that of the hot fluid decreases gradually [Fig. 5.5 (*b*)]. Since, the temperature of one of the fluids remains constant, it is immaterial whether the two fluids flow in the same or opposite directions.

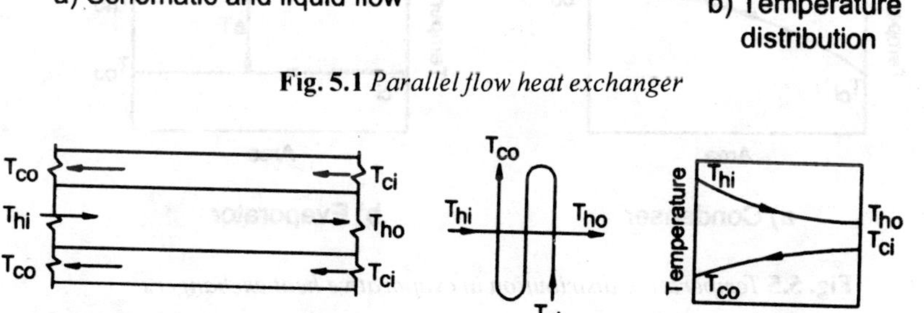

a) Schematic and liquid flow

b) Temperature distribution

Fig. 5.1 *Parallel flow heat exchanger*

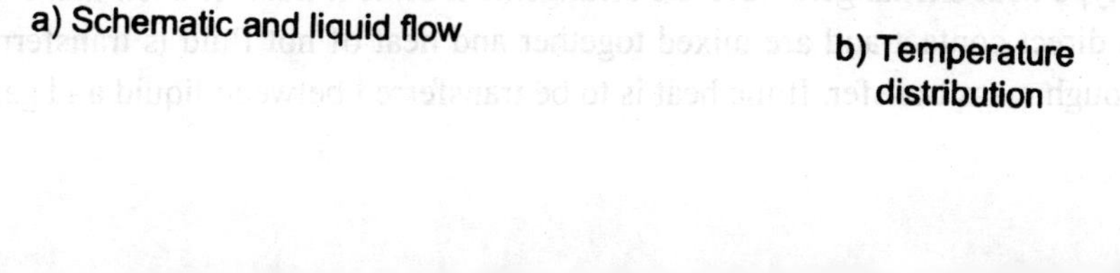

a) Schematic and liquid flow

b) Temperature distribution

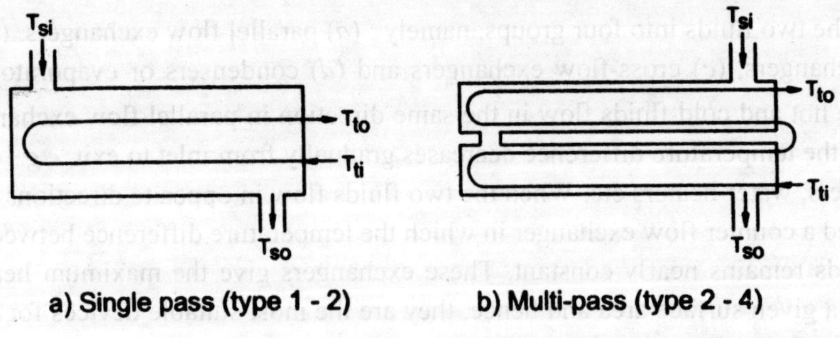

Fig. 5.3 *Single and multi-pass heat exchangers*

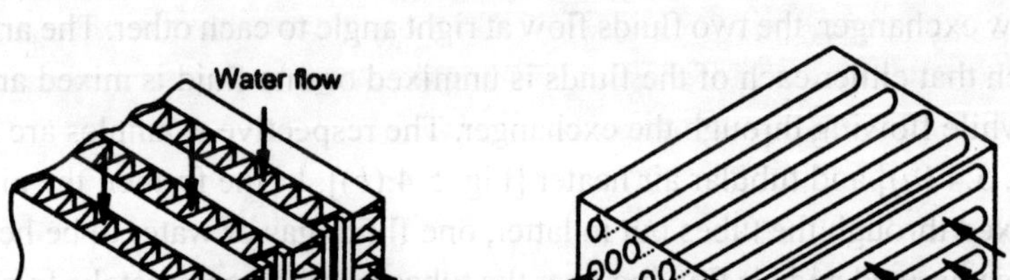

Fig. 5.4 *Cross-flow heat exchanger*

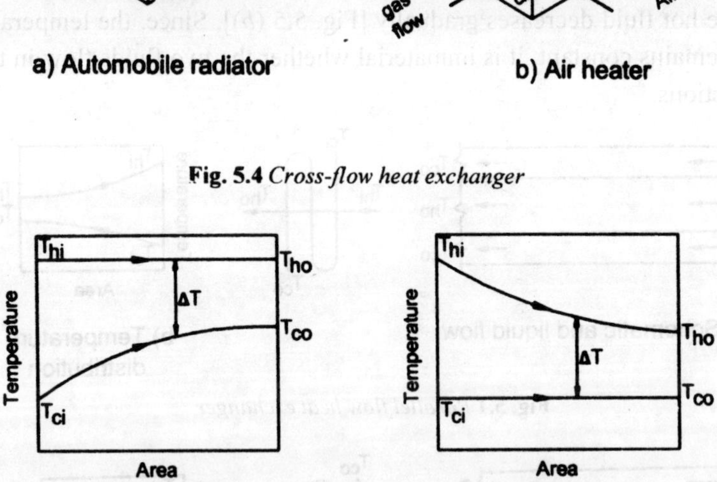

Fig. 5.5 *Temperature distribution in evaporative heat exchangers*

Mixed type heat exchangers have the arrangement of heat transfer such that the two fluids are in direct contact and are mixed together and heat of hot fluid is transferred to cold one through mass transfer. If the heat is to be transferred between liquid and gas, the latter is either bubbled through the former (feed heaters) or the liquid is sprayed in the form of droplets in hot air (spray dryer). Heat is mostly carried by the evaporation of cooled water carried with the air under the influence of relative humidity. Mechanical solutions of this exchanger is not yet developed but done by trial and error only.

5.3 Fluid temperatures in exchangers

The working temperature of both the fluids should either be specified for their inlet temperature or should be fixed based on flow rates and heat transfer considerations. The velocity of liquids and gases should not exceed 7·5 and 30 m/s, respectively. After defining the temperatures, the effectiveness (ratio of fluid temperature rise or drop to the maximum temperature difference) yields which type of flow path is to be used out of parallel, counter or cross flow. While flowing through the heat exchanger the temperature of both the fluids are changing from point to point along the length. Hence, instead of using mean temperature difference $(\Delta T)_m$, the latter can be related with four extreme fluid temperatures which is called as the log mean temperature difference (LMTD) denoted by θ_{lm}. While deriving the expression for this, the following assumptions are made.

(*a*) An overall heat transfer coefficient is constant throughout.

(*b*) Both the fluids flow at a constant flow rate with constant specific heat.

(*c*) During heat transfer, no phase change occurs.

(*d*) Heat losses from the exchanger are negligible.

Considering the inlet and outlet temperature differences, LMTD is given by the following relationship.

$$\theta_{lm} = \frac{(\theta_o - \theta_i)}{\ln(\theta_o/\theta_i)} = \frac{(T_{ho} - T_{co}) - (T_{hi} - T_{ci})}{\ln[(T_{ho} - T_{co})/(T_{hi} - T_{ci})]} \tag{5.1}$$

The arithmetic mean temperature difference (AMTD) denoted as θ_{am} is used for single pass systems instead of θ_{lm} when (θ_i/θ_o) ratio is less than 2·0.

$$\theta_{am} = \frac{(\theta_i + \theta_o)}{2} = \frac{[(T_{ho} - T_{co}) + (T_{hi} - T_{ci})]}{2} \tag{5.2}$$

For the multi-pass and cross flow arrangements a correction factor is used as a multiplying quotient which takes into account the actual flow arrangement of the exchanger. A correction factor is a function of capacity ratio and the temperature ratio. The capacity ratio is defined as the ratio of heat capacity or water equivalent of fluid flowing through the tube to that of the shell. Then,

$$R = \frac{m_t C_{pt}}{(m_s C_{ps})} = \frac{(T_{si} - T_{so})}{(T_{to} - T_{ti})} \tag{5.3}$$

The temperature ratio is defined as,

$$\theta = \frac{(T_{to} - T_{ti})}{(T_{si} - T_{ti})} \tag{5.4}$$

Figure 5.6 shows the values of correction factor for capacity and temperature ratios for commonly used arrangement of multi-pass and cross flow heat exchangers.

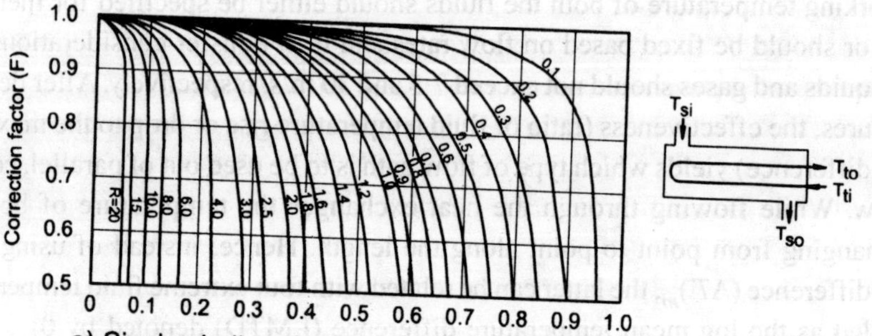

a) Single shell pass and 2, 4, 6..... tube passes

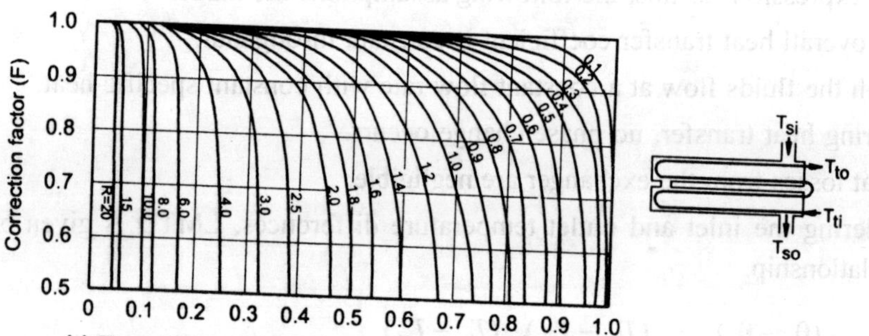

b) Two shell passes and 2, 4, 6..... tube passes

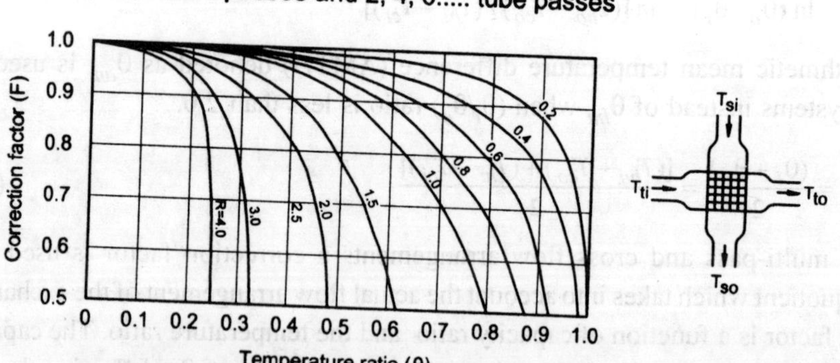

c) Single tube pass cross flow heat excangers

Fig. 5.6 *Correction factor for log mean temperature difference LMTD*

5.4 Overall heat transfer coefficient

It is an important parameter in the design of heat exchangers between the hot and cold fluids. It varies from the composition and geometry of heat flow in the exchanger. The overall heat transfer coefficient of a double tube heat exchanger is given as below based on outer surface of inner pipe.

$$\frac{1}{U_o} = \frac{1}{h_i} \cdot \frac{d_o}{d_i} + \frac{(d_o - d_i)}{(d_o + d_i)} \cdot \frac{d_o}{k} + \frac{1}{h_o} \tag{5.5}$$

Thermal conductivity of some materials is given in Table 5.1. Equation (5.5) can also be expressed as;

$$\frac{1}{U_o} = \frac{1}{h_i} \cdot \frac{r_o}{r_i} + \frac{r_o}{k} \ln\left[\frac{r_o}{r_i}\right] + \frac{1}{h_o} \tag{5.6}$$

Table 5.1 Thermal conductivity of some metals

Metal	*Temperature, °C*	*Conductivity, W/m °C*
Aluminium	0	202
	100	206
Brass (70 Cu, 30 Zn)	0	97
	100	104
	400	116
Copper	0	388
	100	378
Nickel	0	62
	212	59
Cupro-nickel (10% Ni)	0-100	45
Stainless steel (18/8)	0-100	16
Steel	0	45
	100	45
	600	36
Titanium	0-100	16

During operation the salts and other solid materials are deposited on the heat transfer surfaces of the exchanger, called as scaling and fouling of the surface. An additional resistance is added in heat transfer resulting in greatly reduced exchanger effectiveness. To take care of this fact, a "fouling factor" should be taken into account to add the allowance for scaling and fouling of heat transfer surfaces. Considering the scale heat transfer coefficients of inner and outer surface of inner pipe, the total thermal resistance on the basis of outer surface of pipe is given by,

$$\frac{1}{U_o} = \frac{1}{h_i} \cdot \frac{r_o}{r_i} + \frac{r_o}{r_i} \cdot \frac{1}{h_{si}} + \frac{r_o}{k} \cdot \ln\left[\frac{r_o}{r_i}\right] + \frac{1}{h_{so}} + \frac{1}{h_o} \tag{5.7}$$

Table 5.2 gives the scale heat transfer coefficients of fluids of common applications and Table 5.3 shows the overall heat transfer coefficients of typical arrangements for shell and tube equipments. Inverse of scale heat transfer coefficient is termed as the fouling factor.

Table 5.2 Scale heat transfer coefficient for different materials

	Fluid	*Coefficient, W/m²C*
1.	Distilled water, liquid gasoline	11600
2.	Sea water, treated boiler feed water, diesel exhaust gas	5800
3.	Well water, refrigerant vapours	2900
4.	River water	3000-12000
5.	Cooling water in towers	3000-6000
6.	Steam condensate	1500-5000
7.	Steam (oil free)	4000-10000
8.	Steam (oil laden)	2000-5000
9.	Refrigerated brine	3000-5000
10.	Air and industrial gases	5000-10000
11.	Flue gases	2000-5000
12.	Organic vapours	5000
13.	Organic liquids	5000
14.	Heat transfer fluids	5000
15.	Aqueous salt solution	3000-5000

5.5 Double pipe heat exchanger

A double pipe or tubular heat exchanger is of various forms which is assembled easily providing a low-cost heating surface. The most commonly used form is a double pipe exchanger as shown in Fig. 5.7. In this case, the product flows through the inner tube and the heating fluid, generally water, flows through the outer annular space, called jacket, as the co-current or counter-current flow. It is assembled in 4, 5 and 7 m effective lengths, since longer inner pipe sags causing poor hot fluid flow distribution in the jacket. This exchanger is mostly employed where less than 4·0 m^2 heat transfer area is required. It requires larger space for installation providing limited heat transfer area and has an excessive periodic maintenance. The fluids are circulated by pumps and heat flow takes place from hot fluid to inner pipe wall and finally to the inner cold product through forced convection. The design of such exchangers require the knowledge of heat transfer coefficient between the fluids and pipe wall between them.

Table 5.3 Approximate overall heat transfer coefficients for shell and tube equipments

Hot fluid	*Cold fluid*	*Coefficient* (U_o), W/m^2 *C*
Heat exchangers (no phase change)		
Water	Water	850-1700
Organic solvents	Water	280-850
Light oils	Water	340-910
Heavy oils	Water	60-280
Gases	Water	17-280
Organic solvents	Tight oils	115-400
Water	Brine	570-1140
Organic solvents	Brine	170-510
Gases	Brine	20-280
Organic solvents	Organic solvents	115-340
Heavy oils	Heavy oils	45-280
Light oils	Heavy oils	100-400
Gases	Gases	10-50
Evaporators		
Steam	Water	2000-4250
Steam	Organic solvents	570-1140
Steam	Light oils	455-1020
Steam	Heavy oils (vacuum)	140-425
Water	Refrigerants	425-850
Organic solvents	Refrigerants	170-570
Condensers		
Steam (pressure)	Water	2000-4250
Steam (vacuum)	Water	1700-3400
Aqueous vapours	Water	1000-1500
Organic vapours	Water	700-1000
Vacuum condensers	Water	200-500

Contd.

Hot fluid	*Cold fluid*	*Coefficient* (U_o), $W/m^2\ C$
Coolers		
Organic solvents	Water	250-750
Light oils	Water	350-900
Heavy oils	Water	60-300
Gases	Water	20-300
Organic solvents	Brine	150-500
Water	Brine	600-1200
Gases	Brine	15-250
Heaters		
Steam	Water	2000-4250
Steam	Organic solvents	570-1140
Steam	Light oils	300-850
Steam	Heavy oils	60-455
Steam	Gases	30-300
Flue gases	Steam	30-100

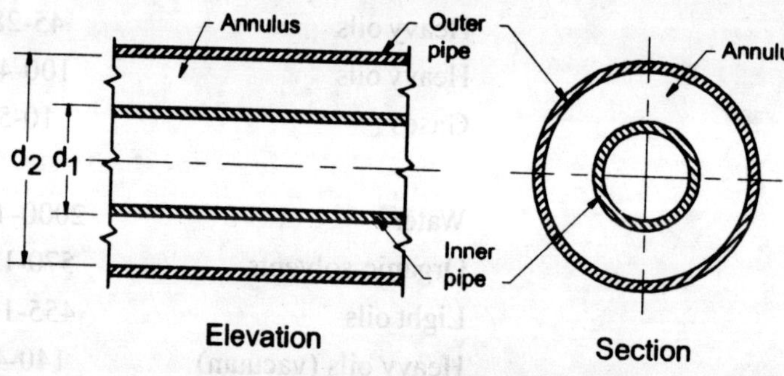

Fig. 5.7 *Double pipe heat exchanger*

Fluid properties are grouped in the form of dimensionless groups as below.

$$N_{Nu} = \text{Nusselt number} = \frac{h_i\, d_e}{k_f} \qquad [(5.8\ (a)]$$

N_{Re} = Reynolds number

$$= \frac{\mu_t d_e \rho}{\mu} = \frac{G_t d_e}{\mu} = \frac{u_t d_e}{\nu} \qquad [(5.8\ (b)]$$

$$N_{Pr} = \text{Prandtl number} = \frac{C_p\, \mu}{k_f} \qquad [(5.8\ (c)]$$

N_{St} = Stanton number

$$= \frac{h_i}{u_t C_p \rho} = \frac{N_{Nu}}{(N_{Re} \cdot N_{Pr})} \qquad [(5.8\ (d)]$$

Equivalent or hydraulic diameter defined as the ratio of four times the cross sectional area of flow to the wetted perimeter given by,

$$d_e = \frac{4A}{P} = \frac{4\pi\left(d_2^2 - d_1^2\right)}{(4\pi d_1)} = \frac{\left(d_2^2 - d_1^2\right)}{d_1} \qquad (5.9)$$

The flow having the Reynolds number less than 2100 in a conduit or tube of uniform cross section will be laminar and more than 4000 will be turbulent in nature. The convective heat transfer coefficient for viscous fluids can be determined by the formula as proposed by Sieder and Tate (1936).

$$N_{Nu} = 0 \cdot 027\ N_{Re}^{0\cdot 8}\ N_{Pr}^{0\cdot 33} \left(\frac{\mu}{\mu_w}\right)^{0\cdot 14} \qquad (5.10)$$

and for laminar flow,

$$N_{Nu} = 1 \cdot 86 \left(\frac{N_{Re}\ N_{Pr} d_e}{L}\right)^{0\cdot 33} \left(\frac{\mu}{\mu_w}\right)^{0\cdot 14} \qquad (5.11)$$

If $N_{Nu} < 3{\cdot}5$, its value should be considered as 3·5. In laminar flow the tube length has a marked effect on heat transfer rate for $L/d_o < 500$. The transition region between laminar and fully developed turbulent flow should be avoided for design. If, it is necessary, the coefficient should be assessed by both the equations (5.10) and (5.11) and the least value should be considered.

Viscosity correction factor should be considered by first computing the coefficient without correction and then using the following relationship for the determination of tube wall temperature.

$$h_i\,(T_w - T_p) = U_o\,(T_p - T_j) \qquad (5.12)$$

Though equation (5.10) and (5.11) can be used for water, Eagle and Ferguson (1930) proposed an accurate relationship specially for water taking the physical properties in the correlations.

$$h_i = \frac{4200\,(1 \cdot 35 + 0 \cdot 02\ t_t)\ u_t^{0\cdot 8}}{d_i^{0\cdot 20}} \qquad (5.13)$$

Pressure drop through exchanger tubes is given by

$$\Delta p = w\left[\frac{4f' L e u_t^2}{2gd'_e}\right] \tag{5.14}$$

where, $$d_e' = \frac{4\pi(d_2^2 - d_1^2)}{4\pi(d_2 + d_1)} = d_2 - d_1 \tag{5.14a}$$

The dimensionless friction factor can be obtained from Fig. 5.8.

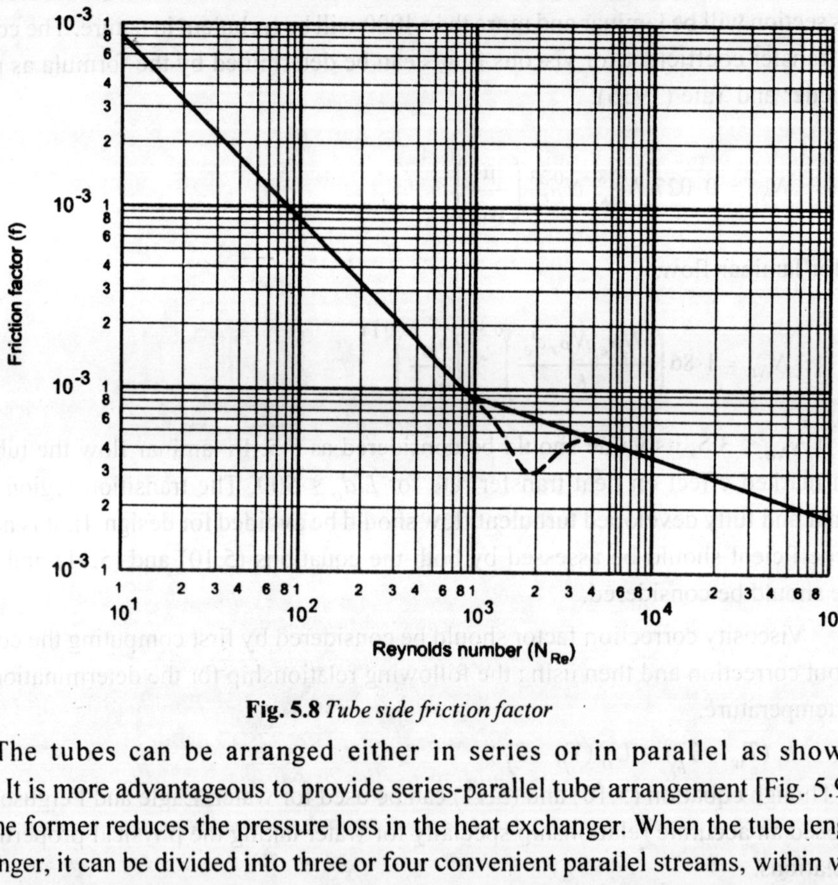

Fig. 5.8 *Tube side friction factor*

The tubes can be arranged either in series or in parallel as shown in Fig. 5.9. It is more advantageous to provide series-parallel tube arrangement [Fig. 5.9 (*b*)] in that the former reduces the pressure loss in the heat exchanger. When the tube length is much longer, it can be divided into three or four convenient parallel streams, within which the fluid can flow in series in case of larger services. When several pipe exchangers are connected in series, the length in the above equation is the total length of the pipe for entire path.

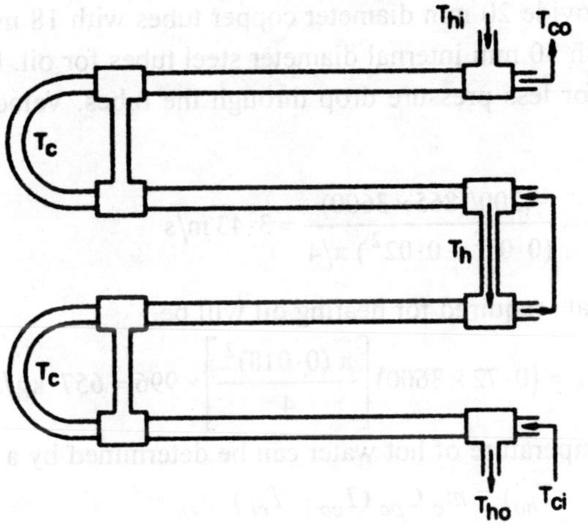

(*a*) Series arrangement

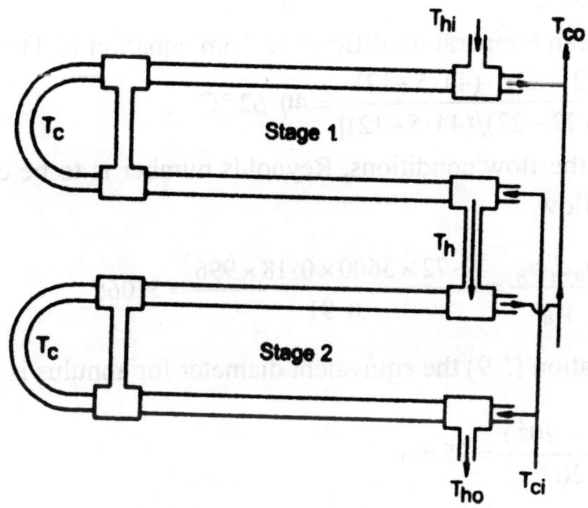

(*b*) Series-parallel arrangement

Fig. 5.9 *Flow arrangement of tubular heat exchanger*

Example 5.1 : *Design a double pipe tubular counter flow heat exchanger (tube length 3·0 m) for heating oil @ 4200 kg/h from 12 to 22 °C using copper tube side hot water available at 72 °C with the fluid properties as below.*

	Property	*Unit*	*Oil*	*Water*
1.	*Thermal conductivity*	*W/mK*	*0·14*	*0·66*
2.	*Bulk density*	*kg/m³*	*865*	*996*
3.	*Specific heat*	*kJ/kg °C*	*1·798*	*4·18*
4.	*Thermal diffusivity*	*m²/s*	*7·3(10)* $^{-6}$	*4·02 (10)* $^{-7}$
5.	*Viscosity*	*kg/hm*	*48*	*0·91*

Solution. Provide 20 mm diameter copper tubes with 18 mm inside diameter for water concentric with 30 mm internal diameter steel tubes for oil. Let the velocity of hot water be 0·72 m/s for less pressure drop through the tubes. Velocity of oil through the annulus will be,

$$u_c = \frac{m_c/\rho_c}{A_c} = \frac{4200/(865 \times 3600)}{(0 \cdot 03^2 - 0 \cdot 02^2)\,\pi/4} = 3 \cdot 43 \text{ m/s}$$

Water flow rate required for heating oil will be,

$$m_h = u_h A_h \rho_h = (0 \cdot 72 \times 3600)\left[\frac{\pi\,(0 \cdot 018)^2}{4}\right] \times 996 = 657 \text{ kg/h}$$

An outlet temperature of hot water can be determined by a heat balance as,

$$m_h\, C_{ph}\,(T_{hi} - T_{ho}) = m_c\, C_{pc}\,(T_{co} - T_{ci})$$

or $\quad 657 \times 4{\cdot}18\,(72 - T_{ho}) = 4200 \times 1{\cdot}798\,(22 - 12)$

then $\quad T_{ho} = 44{\cdot}5\ ^\circ\text{C}$

The log mean temperature difference from equation (5.1) will be,

$$\theta_{lm} = \frac{(72-22)-(44 \cdot 5-12)}{\ln[(72-22)/(44 \cdot 5-12)]} = 40 \cdot 62\ ^\circ\text{C}$$

To decide the flow conditions, Reynolds number is to be calculated for both the fluids. For water flow,

$$N_{Reh} = \frac{u_h d_i \rho_h}{\mu_h} = \frac{0 \cdot 72 \times 3600 \times 0 \cdot 18 \times 996}{0 \cdot 91} = 51065$$

Using equation (5.9) the equivalent diameter for annulus is,

$$d_e = \frac{(30^2 - 20^2)}{20} = 25 \text{ mm}$$

Then the Reynolds number for oil will be,

$$N_{Rec} = \frac{u_c\, d_e\, \rho_c}{\mu_o} = \frac{3 \cdot 435 \times 3600 \times 0 \cdot 025 \times 865}{48} = 5571$$

Since, the Reynolds numbers for both the fluid flows is larger than 4000, the flows will be turbulent. The Prandtl number for water flow will be,

$$N_{Prh} = \frac{C_{ph}\, \mu_h}{k_h} = \frac{4 \cdot 18 \times 0 \cdot 91}{(0 \cdot 66 \times 3 \cdot 6)} = 1 \cdot 60$$

From equation (5.10),

$$N_{Nu} = 0 \cdot 027\, N_{Re}^{0.8}\, N_{Pr}^{0.33} = \frac{h_i\, d_i}{k_h}$$

and $h_i = 0{\cdot}027\,(51065)^{0.8}\,(1{\cdot}6)^{0.33}\left(\dfrac{0{\cdot}66}{0{\cdot}018}\right)$

$= 6753\ \text{W/m}^2\text{C}$

The Prandtl number for oil flow will be,

$$N_{Prc} = \frac{C_{pc}\,\mu_c}{k_c} = \frac{1{\cdot}798 \times 48}{(0{\cdot}14 \times 3{\cdot}6)} = 171{\cdot}24$$

then $h_o = 0{\cdot}027\,(5571)^{0.8}\,(171{\cdot}24)^{0.33}\left(\dfrac{0{\cdot}14}{0{\cdot}025}\right)$

$= 819{\cdot}2\ \text{W/m}^2\text{C}$

Assuming the thermal conductivity of copper as 378 W/mK (Table 5.1) and neglecting fouling factor for water and oil, the overall heat transfer can be determined as below.

$$\frac{1}{U_o} = \frac{1}{6753} \times \left(\frac{0{\cdot}01}{0{\cdot}009}\right) + \frac{0{\cdot}01}{378} \times \ln\left(\frac{0{\cdot}01}{0{\cdot}009}\right) + \frac{1}{819{\cdot}2}$$

$= 0{\cdot}001388$

and $U_o = 720{\cdot}45\ \text{W/m}^2\ \text{C}$

The heat transfer rate required is,

$$q = \frac{75516 \times 1000}{3600} = 20977\ \text{W}$$

Then the number of tubes each of 3·0 m length will be,

$$n = \frac{q}{(U_o \theta_{lm} \pi d_i L)} = \frac{20977}{(720{\cdot}45 \times 40{\cdot}62\pi \times 0{\cdot}02 \times 3{\cdot}00)} = 3{\cdot}8 \text{ say } 4$$

Therefore, provide total 4 tubes in series as shown in Fig. [5.9 (*a*)]. Frictional pressure loss will be $f_h = 0{\cdot}0028$ and $f_c = 0{\cdot}0065$ for $N_{Re} = 51065$ and 2228, respectively from Fig. 5.8. Then,

$$\Delta p_h = \left(\frac{996 \times 9{\cdot}81}{(10)^6}\right) \frac{4 \times 0{\cdot}0028 \times 12 \times (0{\cdot}72)^2}{2 \times 9{\cdot}81 \times 0{\cdot}018} = 0{\cdot}002\ \text{N/mm}^2$$

and $$\Delta p_e = \left(\frac{865 \times 9{\cdot}81}{(10^6}\right) \frac{4 \times 0{\cdot}0065 \times 12 \times (3{\cdot}435)^2}{2 \times 9{\cdot}81 \times 0{\cdot}025} = 0{\cdot}064\ \text{N/mm}^2$$

These pressure drops are small and hence, the design is acceptable.

Example 5.2 : *A tubular heat exchanger has an overall heat transfer coefficient of 350 W/m^{2}K with effective heat transfer area of 14 m^2. Air at 110 °C is to be cooled in this exchanger @ 2·2 kg/s using water available at 16 °C @ 2·1 kg/s. Compute the outlet temperatures of both the air and water streams.*

Solution. The outlet temperatures of both the streams are not given. Hence, the problem can be solved by trail and error by assuming one of the outlet temperatures and coping up the actual rate of heat transfer with the assessed rate. Considering the heat balance,

$$m_h C_{ph} (T_{hi} - T_{ho}) = m_c C_{pc} (T_{co} - T_{ci})$$

then $2{\cdot}2 \times 1{\cdot}0 \,(110 - T_{ho}) = 2{\cdot}1 \times 4{\cdot}19 \,(T_{co} - 16)$

or $(110 - T_{ho}) = 4\,(T_{co} - 16)$

or $T_{ho} + 4T_{co} = 174$

Assuming water outlet temperature as $T_{co} = 30$ °C, then

$T_{ho} + 4 \times 30 = 174$

or $T_{ho} = 54$ °C

Then, log mean temperature difference is given by equation (5.1) as below.

$$\theta_{lm} = \frac{(110-30)-(54-16)}{\ln[(110-30)/(54-16)]} = 56{\cdot}42 \text{ °C}$$

The rate of heat transfer is,

$$q = U_o A_c \theta_{lm} = \left(\frac{350}{1000}\right) \times 14 \times 56{\cdot}42 = 276{\cdot}46 \text{ kJ/s}$$

From heat capacity of water, the rate of heat transfer is,

$$q = m_c C_{pc} (T_{co} - T_{ci}) = 2{\cdot}1 \times 4{\cdot}19\,(30 - 16) = 123{\cdot}2 \text{ kJ/s}$$

Since, this value does not check with the value determined on the basis of $T_{co} = 30$ °C, the latter may be increased. Let $T_{co} = 35$ °C.

then $q = 2{\cdot}1 \times 4{\cdot}19\,(35 - 16) = 167{\cdot}2$ kJ/s

and $T_{ho} + 4 \times 35 = 174$

or $T_{ho} = 34$ °C, $\theta_{lm} = 39{\cdot}94$ °C and $q = 195{\cdot}71$ kJ/s

Still T_{co} should be slightly higher, say $T_{co} = 36$ °C, then,

$q = 2{\cdot}1 \times 4{\cdot}19\,(36 - 16) = 176$ kJ/s

and $T_{ho} + 4 \times 36 = 174$

or $T_{ho} = 30$ °C, $\theta_{lm} = 36{\cdot}03$ and $q = 176{\cdot}55$ kJ/s

In this case, the values of heat transfer rate are very near and hence, air and water temperatures are 30 and 36°C, respectively.

5.6 Shell and tube heat exchanger

It is usually employed where a large heat transfer area is required and is preferred due to less number of the leakage points, less space requirement, easily cleaned, low-cost and tolerating wide range of pressure drops. Many construction materials can be used for the fabrication of shell and tube heat exchanger. It consists of a bundle of tubes, offering a heat transfer surface, through which one of the fluids make the number of passes. Tubes are fixed with the help of tube sheets and are mounted inside the cylindrical shell. The second fluid which is to be condensed or vaporized, flows through the space between tubes and shell. It flows in a zigzag path made by baffles along or across the tubes and transfers heat at increased rates *i.e.*, by increasing the residence time of shell fluid and by improving the heat transfer coefficient. Although, two fluids are physically separated from one another, they are in thermal contact, flowing the heat from hot to cold fluid through tube wall.

5.6.1 Types of heat exchangers

Depending on the mechanical configuration *i.e.*, detailed constructional features and differential thermal expansion between the tube bundle and shell, these types of heat exchangers can be sub-divided in five groups as below.

They are constructed with the following parts :

1. shell, 2. head, 3. flange, 4. nozzle, 5. baffle, 6. tubes, 7. gasket, 8. floating head, 9. channel, 10. channel cover, 11. pass partition, 12. packed gland, 13. expansion below, 14. lifting lug, 15. fixed tube sheet, 16. floating tube sheet, 17. tie rod and spacers, 18. vent connection, 19. drain connection, 20. instrument connection and 21. supports

(*a*) ***Fixed tube-sheet heat exchanger :*** It is the most widely used exchanger type where the temperature difference and shell pressure are less than 80 °C and 0·8 N/mm^2, respectively and shell side fluid is clean and non-fouling. Straight tubes are welded to two-tube sheets, one at each end which facilitates easy cleaning on the inside (Fig. 5.10). Shell is also welded to tube sheets or latter serve as flanges for gasketed joint for bonnets. This makes cleaning and inspection of outside of tubes but there is no possibility of intermixing both the fluids. For the differential thermal expansion of tubes and shell, bellows are provided on shell. Tube side fluid enters and exits through the nozzles fitted on side channel cover.

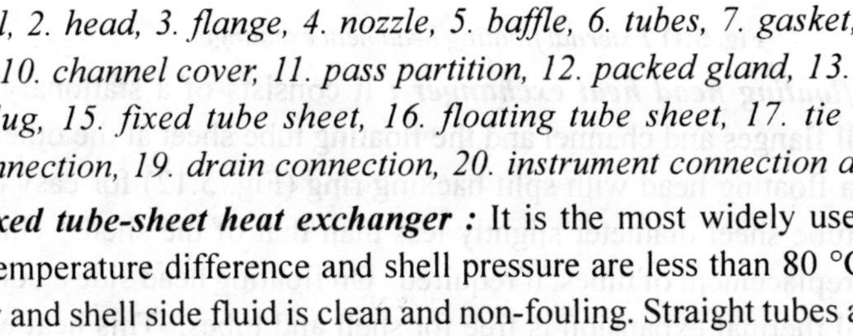

Fig. 5.10 *Fixed tube sheet heat exchanger*

(*b*) ***Outside packed floating head heat exchanger :*** This is a single-pass heat exchanger (Fig. 5.11) rarely used as a vertical condenser with vapour on tube side for restricted temperature and pressure. Shell made in two parts is joined by a stuffing box and gland type joint. A flange is welded on other end of main shell on which a tube sheet is clamped between it and bonnet flange with gasket on either sides. A flange is provided on the remainder end of short shell to which the second tube sheet is fixed with the help of a snap ring. The channel flange is fixed to the tube sheet by a gasketed flanged joint. When the tubes expand, the short shell along with the channel are free to move.

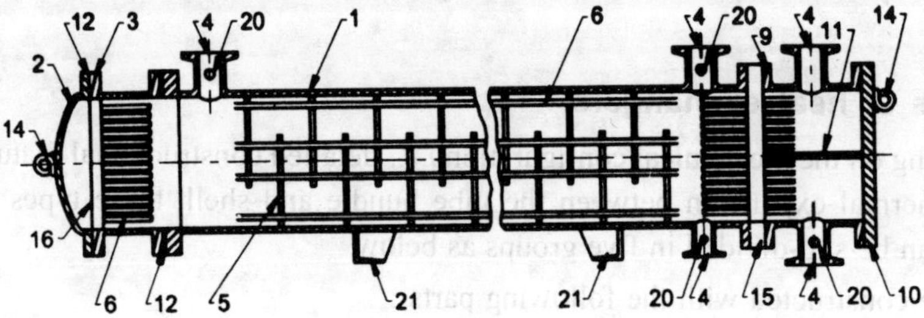

Fig. 5.11 *External floating head heat exchanger*

(*c*) ***Internal floating head heat exchanger :*** It consists of a stationary tube sheet fixed between shell flanges and channel and the floating tube sheet at the other end of the tubes covered by a floating head with split backing ring (Fig. 5.12) for easy dismantling having a floating tube sheet diameter slightly less than that of the shell. This facilitates easy cleaning and replacement of tubes, if required. On floating head side a bonnet is fixed and the differential thermal expansion is free for shell and tubes. This heat exchanger is employed mostly in chemical industries where high temperature and pressure with dirty fluids are used. Because of several gasketed joints, it is costly equipment.

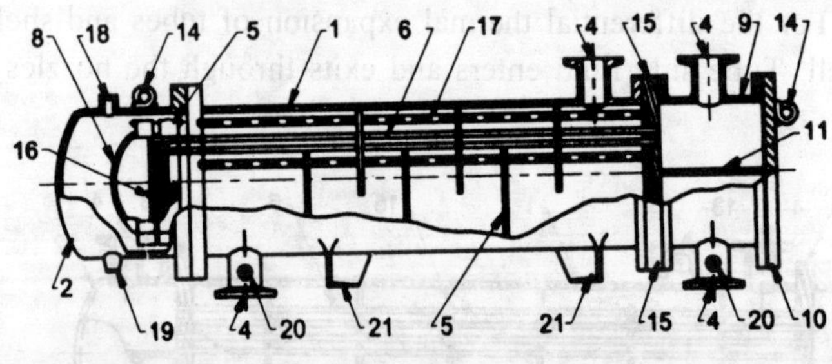

Fig. 5.12 *Internal floating head heat exchanger*

(*d*) ***U-tube heat exchanger :*** The tubes of this exchanger are bent in a hairpin like U-shape (Fig. 5.13) with both the ends fixed to a single stationary tube sheet which facilitates differential thermal expansion between shell and tubes, reducing the number of joints and the cost too. Tube bundle is removable for cleaning and inspection on outside the tubes but the inside of tubes requires chemical cleaning. The exchanger is employed for high temperature and pressure.

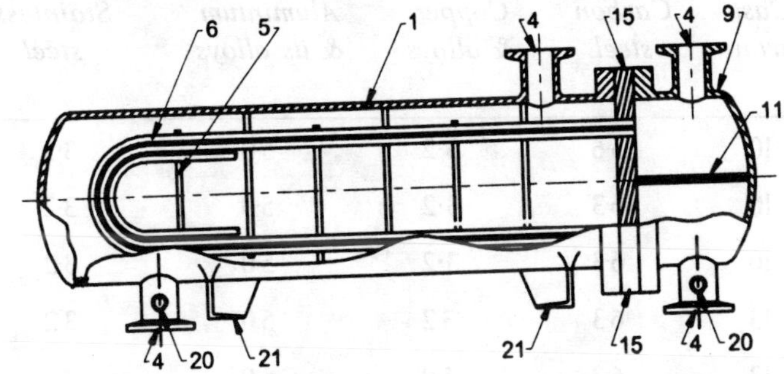

Fig. 5.13 *U-tube heat exchanger*

(*e*) ***Reboiler heat exchanger :*** An internal floating head or U-tube type arrangement is provided in this exchanger with large shell diameter above tube bundle in order to provide a vapour space (Fig. 5.14). For conversion of liquid into vapour, this heat exchanger is provided at the bottom of the distillation column.

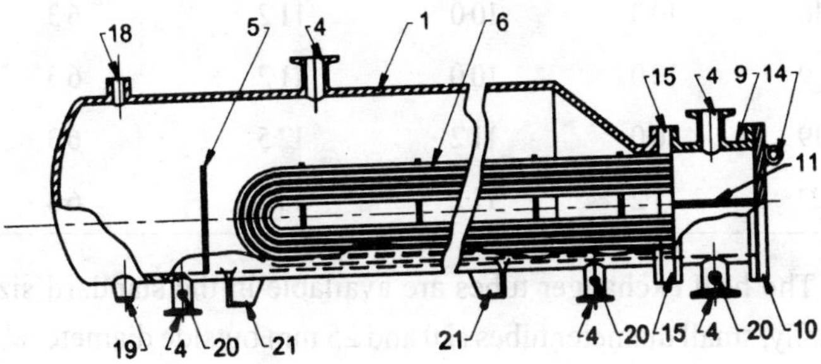

Fig. 5.14 *Kettle reboiler with U-tubes*

5.6.2 Components

(1) *Shell :* It may be fabricated out of a standard steel pipe up to 600 mm diameter or by rolling a' plate. Shell length may be taken four or less times its internal diameter for gas to gas heat exchangers whereas, it may be 4 to 8 times for liquid exchangers. Minimum shell wall thickness may be 5 to 14 mm depending on the nominal diameter and the material of construction (Table 5.4), inclusive of a corrosion allowance. The thickness of shell, nozzle and flanges can be determined by using the equations for pressure vessel design (Chapter 3) considering the joint efficiency.

Table 5.4 Minimum shell thickness (mm) for heat exchangers under normal working conditions

Nominal diameter, mm	*Cast iron*	*Carbon steel*	*Copper & alloys*	*Aluminium & its alloys*	*Stainless steel*	*Nickel*
150	10	5·6	3·2	5·0	3·2	3·2
200	10	6·3	3·2	5·0	3·2	3·2
250	10	6·3	3·2	5·0	3·2	3·2
300	13	6·3	3·2	5·0	3·2	3·2
350	13	6·3	5·0	5·0	3·2	5·0
400	13	6·3	5·0	6·3	3·2	5·0
500	13	8·0	6·3	8·0	3·2	6·3
600	16	8·0	6·3	8·0	5·0	6·3
700	16	10·0	8·3	10·0	5·0	8·0
800	16	10·0	10·0	11·2	6·3	8·0
900	19	10·0	10·0	11·2	6·3	10·0
1000	19	10·0	11·2	12·5	6·3	11·2
1100	22	11·2	11·2	14·0	6·3	11·2

(2) *Tubes* : The heat exchanger tubes are available in the standard sizes (Table 5.5) and lengths. Usually, small diameter tubes (20 and 25 mm outside diameter d_o) are preferred as a given surface can be fitted in a smaller shell diameter and are less expensive. Longer tubes are used for heavy fouling fluids and are easy, for cleaning. Tube lengths incorporated in exchanger as are 0·5, 2.5, 3, 4, 5 and 6 m fixed on the tube sheets by welding, brazing, rolling or screwing. Tubes are arranged either in a equilateral triangular, square or rotated square pattern (Fig. 5.15) with a minimum pitch of 1·25 times the tube outside diameter (S_T = 1·25 d_o). When mechanical cleaning of tube bundle and tube sheets is employed, avoid using a triangular pitch and provide at least 6·5 mm wide tube lanes continuously in a square pitch. The tubes may be arranged for 1 to 16 tube side passes.

The tubes can be bare (plain) or fined, externally or internally depending upon the type of fluids used. The plain tube is used in the normal heat exchanger units. Finned tube exposes increased heat transfer surface.

Table 5.5 Standard dimensions for steel tubes

Material	*Outside tube diameter, mm*	*Wall thickness, mm*
Carbon steel	19·0	1·6, 2·0
	25·4	2·0, 2·6
	31·8	2·0, 2·6, 3·2
	38·0	3·2
Copper and its alloys	6·0	0·6, 0·8
	10·0	0·8, 1·0, 1·2
	12·0	1·0, 1·2
	16·0	1·2, 1·6
	18·0, 20·0	1·2, 1·5, 2·0
	25·0	1·5, 2·0, 2·5
	32·0	1·6, 2·0, 2·5
	40·0	3·0

(3) *Tube sheet* : It is supporting to the tubes and is essentially a flat plate having gasketed joint provision at its periphery. According to the tube pitch holes are drilled (Fig. 5.15) in it to accommodate one tube each. As for as possible there should not be any leakage from the joints. Effective length of tube is found by reducing thickness of the two tube sheets from its total length.

Assuming an equilateral triangular tube pitch [Fig. 5.15 (*b*)], an area occupied by each tube is,

$$a = 0{\cdot}866\ S_T^2 \tag{5.15}$$

Then, the total area covered by tubes is given by,

$$a_n = 0{\cdot}866\ S_T^2\ n \tag{5.16}$$

This area is to be increased for the account of area covered due to the baffles, tie rods etc. by the factor K_b taken as 1 to 1.25 for single tube side pass, 1·18 to 1·43 for double pass and 1·25 to 1·67 for multi-pass exchangers ; to correspond to the cross-sectional area of the shell (A_s). Then,

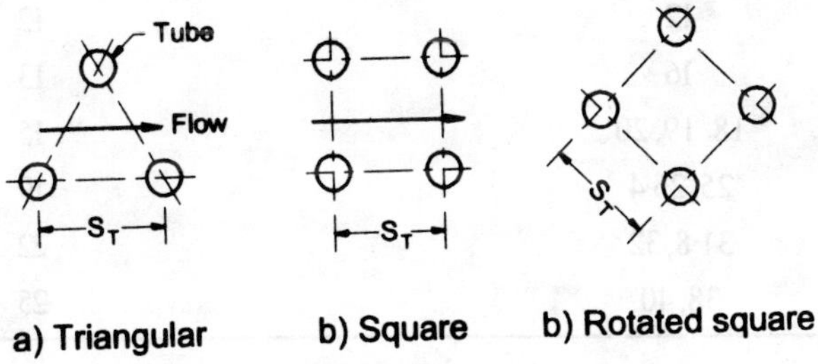

Fig. 5.15 *Tube arrangement on tube sheet*

$$A_s = \frac{\pi D_i^2}{4} = 0{\cdot}866\, S_T^2\, n\, K_b \tag{5.17}$$

where, D_s is the inside diameter of the shell given by,

$$D_i = \left[\left(\frac{4 \times 0\cdot866}{\pi}\right) S_T^2\, n\, K_b\right]^{0\cdot5} = 1{\cdot}05\, S_T\, [n\, K_b]^{0\cdot5} \tag{5.18}$$

For determination of overall tube sheet diameter, gasket thickness should be considered. In case of square pitch $a = S_T^2$ and $a_n = S_T^2\, n$ and expression for shell inside diameter will be,

$$D_i = 1{\cdot}128 S_T\, [n\, K_b]^{0\cdot5} \tag{5.19}$$

Following equation gives the effective tube sheet thickness where corrosion allowance on shell and tube sides should be added and should not be less than given in Table 5.6.

$$t_{ts} = K_s\, G \left[\frac{0\cdot25\, p}{f_a}\right]^{0\cdot5} \tag{5.20}$$

where, K_s is the factor depending on the type of exchanger, taken as 1·0 in most of the types, 1·25 for U-type and for fixed tube sheet exchangers higher value from below is considered for effective thickness.

Tube side pressure,

$$K_s = \left[\frac{(2+K)}{(2+3K)}\right]^{0\cdot5} \tag{5.21}$$

and shell side pressure,

$$K_s = \left[\frac{K}{(2+3K)}\right]^{0\cdot5} \tag{5.22}$$

Table 5.6 Minimum tube sheet thickness for various tube sizes

Tube size, mm (outside diameter)	*Tube sheet thickness, mm*
6	6
10	10
12	12
16	13
18, 19, 20	15
25, 25·4	19
31·8, 32	22
38, 40	25

where

$$K = \frac{E_s\, t_c\, (D_i - t_s)}{E_t\, t_t\, n (d_o - t_t)} \tag{5.23}$$

(4) *Baffles* : In order to induce the velocity and turbulence in shell side fluid, two types of baffles are provided *viz*., longitudinal and transverse. The former type of baffles are extended along the length of shell consisting of flat plates for inducing counter flow of shell and tube side fluids. Several types of transverse baffles are provided in different shapes like segmental, disc and ring, vane and orifice (Fig. 5.16). The most commonly used is the segmental type [Fig. 5.16 (*a*)]. The plates are cut, called window, and are provided with drilled holes and placed to cause the fluid flow up and down. They are supported independent of the tubes with the help of tie rods and positioned by spacers. The tie rods are fixed on tube sheets. Minimum four tie rods of at least 10 mm diameter must be provided. For avoiding the bypass of fluid provide minimum clearance between the shell and baffles and between tubes and baffles.

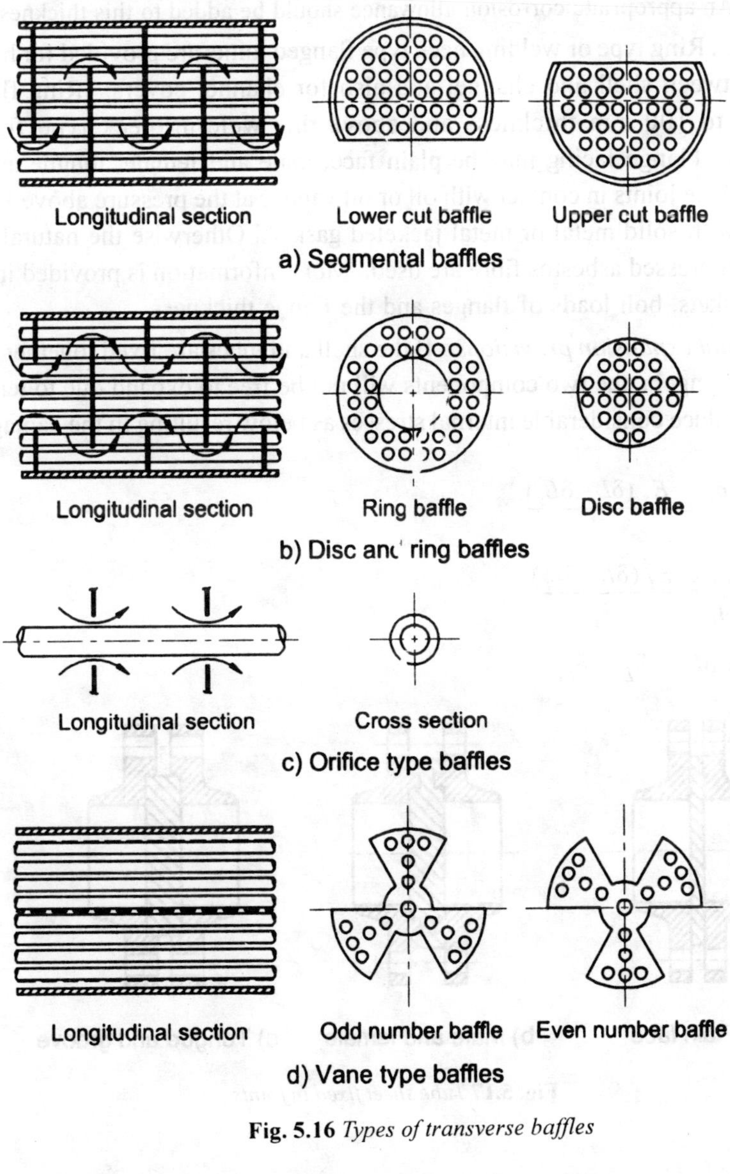

Fig. 5.16 *Types of transverse baffles*

(5) *Channel and its cover* : A channel may be casted or fabricated on piece bonnet type or straight type with a separate bolted cover. Under normal conditions, the minimum cross cover area of channel should be equal to the total flow area of tubes in one pass, but, it should be increased by 30% in case of multi-pass channels operating under severe conditions. The channel wall thickness shall be greater of that determined on the basis of pressure in a cylindrical channel or shell thickness. The channel covers may be dish or dome shaped or flat. The thickness of latter is determined by the following formula.

$$t_s = G\left[\frac{k'\,p}{f_a}\right]^{0.5} \tag{5.24}$$

Value of factor k' is taken as 0·25 for full faced gaskets and 0·3 for narrow faced or ring type gaskets. An appropriate corrosion allowance should be added to this thickness.

(6) *Joints* : Ring type or welding neck type flanged joints are provided for holding the tube sheet between shell and channel and also for channel covers. Ring flanges are preferred up to 100 mm thickness and above this welding neck type flanges are recommended. Flange facing may be plain face, male and female, tongue and groove (Fig. 5.17). All the joints in contact with oil or oil vapour at the pressure above 1·6 N/mm^2 are provided with solid metal or metal jacketed gaskets. Otherwise the natural synthetic rubber and compressed asbestos fibre are used. More information is provided in pressure vessels for gaskets, bolt loads of flanges and the flange thickness.

(7) *Thermal expansion provisions* : If the shell and tubes are fixed together *e.g.* fixed tube sheet exchangers, the two components will not be free to expand due to temperature rise and may induce considerable internal stresses as below resulting in the damage of tube sheets.

$$p_s = \frac{p_t}{A_s} = \frac{E_s\,(\delta L - \delta L_s)}{L_s} \tag{5.25}$$

and
$$p_t = \frac{p_t}{A_t} = \frac{E_t\,(\delta L - \delta L_t)}{L_t} \tag{5.26}$$

Such that $\delta L_s < \delta_L < \delta_{Lt}$.

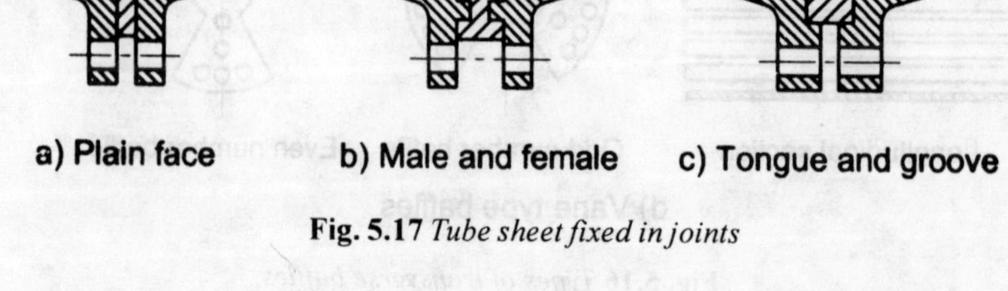

a) Plain face b) Male and female c) Tongue and groove

Fig. 5.17 *Tube sheet fixed in joints*

For the compensation of this differential thermal elongation of shell and tubes, some provisions are made namely, bellows or diaphragms, U-tube or bowed tube arrangement which do not allow to exert the pressures as computed by above equations (5.25) and (5.26). Expansion bellows or diaphragms are flexible which are provided in the shell (Fig. 5.10) and permits constrained expansion or contraction of shell. However, it does not allow the thermal stresses to be zero because of their own rigidity. In case of the U-tube arrangement, each tube is free to expand independently with each other and shell too. In few cases, a short length of tube near the tube sheet is bent creating a flexibility in tubes, called the bowed tube arrangement.

(8) *Nozzles* : Different nozzles are provided on shell and channel of the exchangers for inlets and outlets for shell side and tube side fluids, instrument connection, drainage and ventilation. Nozzles may be integral or welded with shells, shell covers or channels. Channel nozzles can protrude inside the channel except for drainage and ventilation but shell nozzles must be flush with the inside contour of the shell.

(9) *Supports* : The vertical shells are mounted on brackets whereas horizontal shells are supported on two saddles. For fixed tube sheet exchanger, one support is fixed and the other is provided with rollers to facilitate thermal expansion of shell. Supports are separately dealt in Chapter 7.

5.6.3 Design considerations

Thermal design

Numerous parameters are involved in the thermal design of a shell and tube heat exchangers, namely, the nature of heat transfer process, physical properties of fluids, thermal properties of construction materials, arrangement of heat transfer surface and fluid flow rates. The design procedure is based on the use of many performance curves and correlations of different parameters. Since, the exchanger layout cannot be decided unless the heat transfer area is determined, the procedure is on trial and error basis. Few general points should be considered before going for calculations as mentioned below.

(*a*) *Allocation of fluid* : For non-phase changing fluid streams, more corrosive, fouling and high flow rate fluid should be allotted to the tube side. This will reduce the material cost and fouling. Hotter fluid should be circulated through the tubes due to special alloy requirement for higher temperatures and shell surface heat loss at moderate temperatures. Since, high pressure tubes are cheaper than high pressure shell, higher pressure fluid should be passed through the tubes. Higher heat transfer coefficients can be achieved on tube side than on shell side at the same pressure drop and hence, the fluid with lower pressure drop should be allotted to the tube side. More viscous fluid should be passed through the shell for turbulent flow ($N_{Re} > 2000$), but if the flow is laminar ($N_{Re} < 2000$), it should be allotted to the tube side.

(*b*) *Fluid velocities* : High fluid velocity will result in less fouling, high heat transfer coefficients, high pressure drop and may cause erosion. On the other hand, lower velocities may cause settling of the suspended solids of the fluid. For process fluids on tube side, the velocity range should be 1 to 2 m/s and up to 4 m/s for the reduction of fouling and for water, slightly higher range should be used (1·5 to 2·5 m/s), the shell side range being 0·3 to 1·0 m/s. In case of vapours (high molecular weight) the velocity should be 50 to 70 m/s for vacuum, 10 to 30 m/s at atmospheric pressure and 5 to 10 m/s at high pressure.

(*c*) *Pressure drop* : The available pressure drop may vary from a few millibars in vacuum to several bars in the pressure systems. Although the pressure drop is decided by the process conditions, the lower values should be chosen so that the capital and pumping costs should be minimum possible. For liquids with viscosity less than 1·0 mNs/m^2, the pressure drop may be 35 kN/m^2 and for those with viscosity 1 to 10 mNs/m^2, it may be 50 to 70 kN/m^2. In case of gases and vapours the pressure drop should be 0·4 to 0·8 kN/m^2 at high vacuum in exchanger, 0·10 times the absolute pressure for medium vacuum, 0·5 and 0·1 times the system pressure for 0·1 to 0·2 N/mm^2 and 1·0 N/mm^2, respectively.

(*d*) *Fluid stream temperatures* : The difference between the outlet temperature of one stream and an inlet temperature of the other stream should not be less than 20 °C which will require higher area for heat transfer resulting in an uneconomical exchanger.

(*e*) *Fluid physical properties* : Fluid density, viscosity, thermal conductivity, specific heat and latent heat are the physical properties which should be considered for the exchanger design in the form of dimensionless numbers defined in equation (5.8). Physical properties at bulk stream temperatures give satisfactory results when temperature changes are small but may add enormous error at higher temperature changes. In such cases, lower value of the heat transfer coefficient estimated at inlet and outlet should be used for further computations.

Design procedure

Calculations for the thermal design of shell and tube heat exchanger can be made by the semi-analytical procedure.

(1) Assess the inlet and outlet temperatures of both the tube side and shell side fluids with their allocation as described above.

(2) Find out the properties of fluids at their respective bulk temperatures.

(3) Determine the LMTD and required rate of heat flow using equation (5.1) and basic heat transfer equation, respectively.

(4) Knowing the mass flow rate of one fluid, compute for the other fluid by a heat balance and decide/compute the fluid velocities.

(5) Assuming proper tube size and length, compute the overall heat transfer coefficient based on outer surface of tubes by using the fluid and metal thermal properties and equations (5.7) and (5.9) to (5.12). For the determination of heat transfer coefficient for film condensation in case of pure vapours over a plane vertical surface of length. Nusselt (1976) proposed the following relationship.

$$h_o = 0 \cdot 94 \left[\frac{\rho^2 g \, L_v \, k^3}{\mu \, L (T_{sat} - T_w)} \right]^{0 \cdot 25} \tag{5.27}$$

A corresponding equation for horizontal tube is,

$$h_o = 0 \cdot 725 \left[\frac{\rho^2 g \, L_b \, k^3}{\mu \, d_o (T_{sat} - T_w)} \right]^{0 \cdot 25} \tag{5.28}$$

(6) Compute the total outside tube area required for heat transfer, total number of tubes required and tube passes.

$$A_o = \frac{q}{(U_o \, \theta_{lm})} \tag{5.29}$$

$$n = \frac{A_o}{(\pi \, d_o \, L)} \tag{5.30}$$

Then number of tubes per pass will be

$$n_p = \frac{m_t / \rho_t}{u_t \, (\pi/4) \, d_o^2} = \frac{4 m_t}{\rho_t \, u_t \, \pi \, d_o^2} \tag{5.31}$$

and $$N_t = \frac{n}{n_p}$$

(7) Estimate the pressure drop through tubes using the following relationship.

$$\Delta P_t = \frac{N_t \, \rho_t \, u_t^2}{2} \left\{ 8 \, f_t \left[\frac{L}{d_i} \right] \left[\frac{\mu_t}{\mu_w} \right]^{-0 \cdot 25} + 0 \cdot 25 \right\} \tag{5.32}$$

The friction factor is based on the Reynolds number on tube side obtained from Fig. 5.8. Factor 0·25 takes into account the frictional pressure loss due to sudden contraction and expansion at tube inlet and outlet, respectively and also due to the channel friction. The viscosity effect may be neglected if the difference is small. If the estimated value of Δp_t is within limit, the design is acceptable, otherwise revise the number of tubes using the following equation (approximately) and repeat the calculations.

$$n = \frac{0 \cdot 9 \, m_t}{\Delta p_{tm} \, d_o^2} \left\{ \frac{N_t^3}{\rho} \left(8 f_t \left[\frac{L}{d_i} \right] \left[\frac{\mu_t}{\mu_w} \right]^{-0 \cdot 25} + 0 \cdot 25 \right) \right\}^{0 \cdot 5} \tag{5.33}$$

(8) Deciding the tube pitch (triangular or square) compute the tube bundle diameter (using the following equation and shell internal diameter by using equation (5.18)

$$D_b = d_o \left(\frac{n}{k_1} \right)^{(1/n_1)} \tag{5.34}$$

The coefficients k_1 and n_1 depend on the tube pitch and number of tube passes as given in Table 5.7.

(9) Choose a suitable baffle spacing usually $D_s/5$ and then compute the cross flow area. Also calculate the equivalent diameter of the shell.

$$A_s = \frac{(S_T - d_o)}{S_T} B_f D_s \tag{5.35}$$

and $$d_e = \left(\frac{1 \cdot 1}{d_o} \right) \left(S_T^2 - 0 \cdot 917\, d_o^2 \right) \tag{5.36}$$

(10) Estimate the pressure drop through the shell. If it is within limit, accept the design. If greater is the pressure drop adjust the shell side fluid velocity by increasing the baffle spacing in order to reduce it.

Table 5.7 Values of constants to be used in equation (5.34)

	Constant	*Number of passes*		
		1	*2*	*3*
1.	Square pitch			
	k_1	0·215	0·156	0·158
	n_1	2·207	2·291	2·263
2.	Triangular pitch			
	k_1	0·319	0·249	0·175
	n_1	2·142	2·207	2·285

$$\Delta p_s = \left(\frac{8\, f_s u_s^2 \rho_s}{2} \right) \left(\frac{D_i}{D_e} \right) \left(\frac{L}{B_f} \right) = 4\, f_s u_s^2 \rho_s \left(\frac{D_i}{D_e} \right) \left(\frac{L}{B_f} \right) \tag{5.37}$$

Mechanical design

Due to wide application, the shell and tube heat exchangers are standardized for the design and construction. Different codes of practice specify the design procedures and details about the constructional features *viz.* IS : 4503-1967, BS : 3274 and TEMA (Tubular Exchanger Manufacturer's Association). A brief procedure is laid down below for mechanical design of shell and tube heat exchangers.

Shell side parameters

(1) Determine the shell thickness using equation (3.5) and add suitable corrosion allowance so that the minimum thickness is satisfied as given in Table 3.2. Also provide the nozzle thickness using the same equation.

(2) Provide shallow dished or torispherical type head for shell with dish or crown radius $\leq$ the head diameter. Compute the intensification factor first using equation (3.19) and then determine the head thickness using equation (3.18). Provide corrosion allowance suitably to this thickness and choose standard thickness for fabrication.

(3) Design the flanged joint between the shell and tube sheet. For this, select suitable gasket for given operating conditions from Table 3.4, find out respective seating stress and gasket factor from Table (3.5) and decide inner and outer gasket diameters using equation (3.41).

(4) Decide diameter of gasket load reaction and calculate bolt loads under operating and bolting up conditions from equations (3.46) and (3.47), respectively. Determine bolting area with larger of the two loads to the allowable stress in bolts (Table 3.8) and then number of bolts required after selection of suitable bolts from Table 3.7. Assuming suitable hub thickness determine pitch circle diameter using equations (3.44) and (3.43). Adjust the bolt spacing so that both the values are near to each other. Check the gasket prevention from condition given in equation (3.48) and then determine the flange thickness using equations (3.50) and (3.49), with the extra corrosion allowance.

Tube side parameters

1. Determine tube wall thickness as a pressure pipe based on outside diameter using equation (3.6).

2. Calculate the tube sheet thickness using equation (5.20). Equation (5.24) gives channel cover thickness.

3. Design flanged joint, gasket and bolting for tube sheet and channel cover similar to that of the shell side flange. Also compute the flange thickness.

4. Compute the nozzle thickness from equation (3.6) using the nozzle parameters.

Example 5.3 : *Cold water is to be preheated from 16 to 48 °C circulated through 20 mm inside diameter tubes using hot water available at 96 °C and 3·33 kg/s to be cooled to 51 °C on shell side in a shell and tube heat exchanger. If the tube length and velocity of water are to be limited to 3200 mm and 500 mm/s, respectively, propose a suitable design for heat exchanger assuming an overall heat transfer coefficient as 1480 W/m² K, neglecting the metal and fouling resistance factors.*

Solution. Given T_{ho} = 51 °C, T_{hi} = 96 °C, T_{co} = 48 °C, T_{ci} = 16 °C, then,

From equation (5.1),

$$\theta_{lm} = \frac{(96-48)-(51-16)}{\ln\left[(96-48)/(51-16)\right]} = 41{\cdot}16\ °\text{C}$$

Total heat load with hot water will be transferred to cold water. With specific heat of water as 4160 W/kg,

$$q = m_g\ C_{ps}\ \Delta T = 3{\cdot}33 \times 4160\ (96-51) = 623376\ \text{W}$$

Tube side parameters : Total heat transfer area on outside tube surface will be,

$$A_o = \frac{q}{(U_o \theta_{lm})} = \frac{623376}{(1480 \times 41{\cdot}16)} = 10{\cdot}23\ \text{m}^2$$

For this area number of tubes with length of 3200 – 40 = 3160 mm will be,

$$n = \frac{A_o}{(\pi\, d\, L)} = \frac{10{\cdot}23}{(\pi \times 0{\cdot}02 \times 3{\cdot}16)} = 51{\cdot}52\ \textit{i.e.}, 52\ \text{tubes}$$

With the temperature conditions and available heat with hot water, the flow rate for cold water is,

$$m_t = \frac{623376}{[4160\ (48-16)]} = 4{\cdot}68\ \text{kg/s}$$

and cross sectional area of flow will be,

$$A_t = \frac{m_t}{\mu_t \rho_t} = \frac{4{\cdot}68}{(0{\cdot}5 \times 996)} = 9{\cdot}4\ (10)^{-3}\ \text{m}^2$$

and number of tubes per pass will be,

$$n_p = \frac{9{\cdot}4\ (10)^{-3}}{[0{\cdot}25\ \pi (0{\cdot}02)^2]} = 29{\cdot}92 \quad \textit{i.e.}\ 30\ \text{tubes}$$

Hence, two number of tube passes are proposed. Assuming μ_t = 2·764 kg/mh

$$N_{Re} = \frac{u_t\, d_t\, \rho_t}{\mu_t} = \frac{(0{\cdot}5 \times 3600) \times 0{\cdot}02 \times 996}{2{\cdot}764} = 12972$$

For N_{Re} = 12972, f_t = 0·004 from Fig. 5.8 and neglecting (μ_t/μ_w) term in equation (5.32) the pressure drop will be,

$$\Delta p_t = \frac{(2 \times 996\ (0{\cdot}5)^2}{2} \cdot \left[8 \times 0{\cdot}004 \left(\frac{3{\cdot}16}{0{\cdot}02}\right) + 0{\cdot}25\right] = 1321.2\ \text{N/m}^2$$

Since, this pressure drop is very small, design is acceptable.

Shell side parameters : From Table 5.7, k_1 = 0.249 and n_1 = 2.207 for two passes and from equation (5.34) the diameter of tube bundle will be,

D_b = 20 (2 × 30 / 0·249)$^{(1/2 \cdot 207)}$ = 240·05 mm, hence, provide D_b = 240 mm. Providing a square pitch, S_T = 1·25 × 20 = 25 mm and a = $(25)^2$ = 625 mm². Using equation (5.18), the internal diameter of shell will be,

$$D_i = 1 \cdot 05 \times 25\,(2 \times 30 \times 1 \cdot 33)^{0 \cdot 5} = 234 \cdot 5 \text{ mm}$$

also $\quad D_i = 1 \cdot 25\, D_b = 1 \cdot 25 \times 240 = 300$ mm

Hence, provide D_i = 300 mm and baffle pitch as $D_i/4$ = 75 mm

The cross flow area is given by equation (5.35) as,

$$A_s = \left[\frac{(25-20)}{25}\right] \times 300 \times 75 = 4500 \text{ mm}^2$$

Velocity of heating water in shell will be,

$$u_s = \frac{m_s}{(A_s\, \rho_s)} = \frac{3 \cdot 33}{[4500\,(10)^{-6} \times 996]} = 0 \cdot 74 \text{ m/s}$$

Equation (5.36) gives the equivalent diameter,

$$d_e = \left(\frac{1 \cdot 1}{20}\right)(25^2 - 0 \cdot 917 \times 20^2) = 14 \cdot 2 \text{ mm}$$

and $$N_{Re} = \frac{(0 \cdot 74 \times 3600) \times 0 \cdot 0142 \times 996}{1 \cdot 38} = 27302$$

For this, friction factor, f_s = 0·0034 from Fig. 5.8. Then from equation (5.37), the pressure drop through the shell will be,

$$\Delta p_s = 4 \times 0 \cdot 0034 (0 \cdot 74)^2 \times 996 \left(\frac{300}{14 \cdot 2}\right)\left(\frac{3160}{75}\right)$$
$$= 6656 \text{ N/m}^2 = 0 \cdot 007 \text{ kN/m}^2$$

This pressure drop is small and hence, the design is acceptable.

Example 5.4 : *A 1-4 shell and tube heat exchanger is proposed for condensing ammonia vapours at 1·88 N/mm² pressure, flowing at 0·3 kg/s. Water is available at 28 °C and flowing through 1·5 mm thick 25 mm outside diameter 4·80 m long stainless steel tubes arranged in a triangular pitch of 32 mm. Water leaves the tubes at 33 °C. An overall heat transfer coefficient is 1000 W/m² °C, saturation temperature of ammonia vapour is 42 °C and its latent heat of vaporization is 1·09 MJ/kg. It is proposed to fabricate shell using 8 mm thick carbon steel sheet and the velocity of fluids on tube side and shell side is to be restricted to 2·2 and 0·10 m/s, respectively. Determine the number of tubes required and check the shell and tube parameters, assuming tube side pressure as 11·5 N/mm².*

Solution. Given :

T_{ho} = 42 °C, T_{hi} = 28 °C, T_{co} = 42 °C, T_{ci} = 33 °C, m_s = 0·3 kg/s

L_v = 1·09 MJ/kg, p_s = 1·88 × 1·33 = 2·50 N/mm², U_o = 1000 W/m² °C

t_s = 8 mm (carbon steel), f_{as} = 96 N/mm^2, f_{at} = 160 N/mm^2

L_t = 4·80 – 0·05 = 4·75 m, p_t = 11·5 × 1·33 = 15·30 N/mm^2

$$\theta_{lm} = \frac{(42-28)-(42-33)}{\ln[(42-28)/(42-33)]} = 11 \cdot 32 \text{ °C}$$

Total heat load is,

$q = m_s\ L_v$ = 0·3 × 1·09 = 0·327 MW

Tube side parameters : Outside area of tube surface required for heat transfer is,

$$A_o = \frac{0 \cdot 327\ (10)^6}{(1000 \times 11 \cdot 32)} = 28 \cdot 88 \text{ m}^2$$

Number of tubes required will be,

$$n = \frac{28 \cdot 88}{(\pi \times 0 \cdot 025 \times 4 \cdot 75)} = 77 \cdot 41$$

It is proposed to provide 80 tubes so that the number of tubes per pass will be 80/4 = 20. With the available heat load, water flow rate through the tubes will be,

$$m_t = \frac{0 \cdot 327\ (10)^6}{[4160\ (33-28)]} = 15 \cdot 72 \text{ kg/s}$$

Area of flow of water through tubes with d_i = 25 – (2 × 1·5) = 22 mm is,

A_t = 0·25 π (0·022)^2 × 20 = 7·60 (10)$^{-3}$ m^2

then $$u_t = \frac{15 \cdot 72}{[7 \cdot 6\ (10)^{-3} \times 1000]} = 2.07 \text{ m/s}$$

which is within the limit and tube wall thickness required with seamless pipe (j = 1·0) is,

$$t_t = \frac{15 \cdot 3 \times 25}{(2 \times 160 \times 1) + 15 \cdot 3} = 1 \cdot 14 \text{ mm}$$

which is also within the limit.

Shell side parameters : From Table 5.7, k_1 = 0·175 and n_1 = 2·285 for four passes, then,

$$D_b = 25\left(\frac{80}{0 \cdot 175}\right)^{1/2 \cdot 285} = 365 \text{ mm}$$

and D_i = 1·25 × 365 = 456 mm

Hence, provide D_s = 460 mm and $B_s = \frac{460}{4} = 115$ mm

Gross flow area on shell side is, given by equation (5.35)

$$A_s = \left[\frac{(32-25)}{32}\right] \times 480 \times 120 = 12600 \text{ mm}^2$$

then $$u_s = \frac{0 \cdot 3}{[12600\,(10)^{-6} \times 1000]} = 0{\cdot}024 \text{ m/s, } i.e. \text{ within limit}$$

also $$t_s = \frac{2 \cdot 5 \times 480}{(2 \times 96 \times 0 \cdot 85) - 2 \cdot 5} = 7.46 \text{ mm}$$

With 2·5 mm corrosion allowance, 10 mm standard thickness of shell plate will be required instead of 8 mm.

Example 5.5 : *Steam with dryness fraction of 0·92 is to be condensed at 30 °C and 4160 N/m² pressure flowing at 2·25 kg/s. Cooling water is available at 16 °C and the temperature rise has to be limited to 6 °C due to economic reasons. Tubes of 16·5 mm inside and 19 mm outside diameters are available through which the velocity is to be limited to 1·2 m/s in two passes. Latent heat of vaporization is 2400 kJ/kg. Considering an overall heat transfer coefficient of 3300 W/m² °C, suggest a suitable thermal design for the condenser.*

Solution. Total heat load is

$$q = 2{\cdot}25 \times 0{\cdot}92 \times 2400 = 4968 \text{ kW}$$

Water flow rate for this heat load is,

$$m_t = \frac{4968\,(10)^3}{(4160 \times 6)} = 199{\cdot}04 \text{ kg/s}$$

The volumetric flow rate is,

$$V_t = \frac{m_t}{\rho_t} = \frac{199 \cdot 04}{996} = 0{\cdot}20 \text{ m}^3/\text{s}$$

This flow rate is also equal to the product of flow area per pass and water velocity, then,

$$V_t = 0.25\ \pi\ (d_i)^2\ (n/2) \cdot Ut$$

or $$0.20 = 0{\cdot}25\ \pi\ (0{\cdot}0165)^2 \times (n/2) \times 1{\cdot}2$$

or $n = 1559$ *i.e.* provide 1560 tubes, 780 in each pass

$$\theta_{lm} = \frac{(30-16)-(30-24)}{\ln[(30-16)/(30-24)]} = 10{\cdot}72\ °\text{C}$$

Heat transfer area required is,

$$A_o = \frac{4968 \times 1000}{3300 \times 10 \cdot 72} = 140{\cdot}43 \text{ m}^2$$

Length of tube required is,

$$L_t = \frac{140 \cdot 43}{[(0 \cdot 019\ \pi) \times 1560]} = 1 \cdot 51 \text{ m, say } 1 \cdot 80 \text{ m with allowances}$$

Assuming the tube arrangement on square pitch, for two passes, $k_1 = 0 \cdot 156$ and $n_1 = 2 \cdot 291$ from Table 5.7, then tube bundle diameter is,

$$D_b = 0 \cdot 019 \left(\frac{1560}{0 \cdot 156} \right)^{(1/2 \cdot 291)} = 0 \cdot 995 \text{ m}$$

hence, provide $D_b = 1.0$ m

then $\quad D_i = 1 \cdot 0 \times 1 \cdot 10 = 1 \cdot 10$ m

Example 5.6 : *Design a U-tube type shell and tube heat exchanger for heating oil from 12 to 90 °C to be supplied for firing the furnace on horizontal tube side @ 12000 kg/h at 0·28 N/mm² pressure. Steam is available at 130 °C and leaving the shell at 110 °C. Fluid properties at respective bulk temperatures are as below.*

	Property	*Units*	*Oil*	*Steam*
1.	*Bulk density*	*kg/m³*	*910*	*970*
2.	*Viscosity*	*kg/hm*	*55*	*0·97*
3.	*Specific heat*	*kJ/kg °C*	*1·97*	*4·20*
4.	*Thermal conductivity*	*W/m °C*	*0·15*	*0·75*

Solution. *Thermal design :* Outlet and inlet temperatures of hot fluid are 110 and 130 °C and cold fluids are 90 and 12 °C, respectively. Then, using equation (5.1),

$$\theta_{lm} = \frac{(110-90)-(130-12)}{\ln[(110-90)/(130-12)]}$$

$$= 55 \cdot 21 \text{ °C}$$

The required rate of heat transfer will be,

$$q = m_c\, C_{pc}\, (T_{ho} - T_{ci})$$

$$= \frac{12000 \times 1 \cdot 97\,(90-12)}{3 \cdot 6}$$

$$= 512200 \text{ W}$$

Steam requirement for heating the oil can be determined by a heat balance as,

$$m_h\, C_{ph}\, (T_{hi} - T_{ho}) = m_c\, C_{pc}\, (T_{co} - T_{ci})$$

Then $\quad m_h = m_c \dfrac{C_{pc}}{C_{ph}} \cdot \dfrac{(T_{co} - T_{ci})}{(T_{hi} - T_{ho})}$

$$= 12000 \times \frac{1 \cdot 97}{4 \cdot 20} \times \frac{(90-12)}{(130-110)}$$

$$= 21952 \text{ kg/h}$$

Tube side parameters : Providing 20 mm copper tubes with 1·6 mm wall thickness, the internal diameter is d_i = 16·8 mm. Using 6·0 m tube length an effective length L = 5·96 m with tube sheets. In order to get low pressure drop and laminar flow, let, the oil flow velocity be 0·05 m/s. Let, μ_w = 13·2 kg/hm at bulk temperature.

The tube side heat transfer coefficient can be determined by using equation (5.11) which gives,

$$\frac{h_i d_i}{k_i} = 1\cdot 86\left[\frac{u_t d_i \rho_t}{\mu_t}\cdot\frac{C_{pt}\mu_t}{k_t}\cdot\frac{d_i}{L}\right]^{0\cdot 33}\left[\frac{\mu_t}{\mu_w}\right]^{0\cdot 14}$$

or $$h_i = \frac{1\cdot 86\times 0\cdot 15}{0\cdot 0168}\left[\frac{0\cdot 05\times 3600\times 0\cdot 0168\times 910}{55}\times\frac{1\cdot 97\times 55}{0\cdot 15\times 3\cdot 6}\times\frac{0\cdot 0168}{5\cdot 96}\right]^{0\cdot 33}\left[\frac{55}{13\cdot 2}\right]^{0\cdot 14}$$

$$= 61\cdot 10 \text{ W/m}^2\,°\text{C}$$

The heat transfer coefficient for vapour condensing over horizontal tubes is given by equation (5.28). Latent heat of vaporization for steam at 130 °C is,

$$L_V = 2174 \text{ kJ/kg}$$

$$h_o = 0\cdot 725\left[\frac{\rho_s^2\, g\, L_V\, k^3}{\mu_s\, d_o\,(t_{si} - t_{so})}\right]^{0\cdot 25}$$

$$= 0\cdot 725\left[\frac{(970)^2(9\cdot 81\times 3600^2)\times 2174\,(0\cdot 75\times 3\cdot 6)^3}{0\cdot 97\times 0\cdot 02\,(130-110)}\right]^{0\cdot 25}$$

$$= 43694 \text{ kJ/hm}^2\,°\text{C}$$

$$= 12137 \text{ W/m}^2\,°\text{C}$$

For copper tubes the thermal conductivity is 378 W/m°C from Table 5.1, h_{si} = 1160 W/m²°C and h_{so} = 7000 W/m²C. Then, using equation (5.7) the overall heat transfer coefficient can be determined as,

$$\frac{1}{U_o} = \frac{1}{61\cdot 10}\cdot\frac{0\cdot 02}{0\cdot 0168}+\frac{1}{1160}\cdot\frac{0\cdot 02}{0\cdot 0168}+\frac{0\cdot 02}{2\times 378}\ln\left[\frac{0\cdot 02}{0\cdot 0168}\right]+\frac{1}{7000}+\frac{1}{12137}$$

$$= 0\cdot 0254$$

or $$U_o = 39\cdot 21 \text{ W/m}^2\text{K}$$

Using basic heat transfer equation, the total heat transfer area on outside of the tube from equation (5.29) will be,

$$A_o = \frac{435370}{39\cdot 21\times 55\cdot 21} = 236.6 \text{ m}^2$$

For this area, number of 20 mm diameter tubes from equation (5.30) will be,

$$n = \frac{236\cdot 60}{(\pi \times 0\cdot 02 \times 5\cdot 96)} = 631\cdot 8 \text{ tubes}$$

Also, the number of tubes per pass from equation (5.31) is,

$$n_p = \frac{4 \times 12000}{910 \times (0\cdot 05 \times 3600)\ \pi\ ((0\cdot 0168)^2} = 330$$

Thus, two number of tube passes will be required.

The Reynolds number on tube side is,

$$N_{Re} = \frac{u_t\, d_i\, \rho_t}{\mu_t} = \frac{(0\cdot 05 \times 3600) \times 0\cdot 0168 \times 910}{55} = 50$$

The friction coefficient for N_{Re} = 50, from Fig. 5.8, is 0.16, then from equation (5.32), the pressure drop through the tubes will be,

$$\Delta p_t = \left[\frac{2 \times 910\,(0\cdot 05)^2}{2}\right] \times \left[8 \times 0\cdot 16 \left(\frac{5\cdot 96}{0\cdot 0168}\right)\left(\frac{55}{13\cdot 2}\right)^{-0\cdot 25} + 0\cdot 25\right]$$

$$= 723\cdot 64 \text{ N/m}^2$$

Since, the pressure drop is low, there is a possibility to increase the number of tube passes and the design is acceptable.

Shell side parameters : Using a triangular pitch the tube bundle diameter (D_b) is given by equation (5.34). For two passes, $k_1 = 0\cdot 249$ and $n_1 = 2\cdot 207$, from Table 5.7. Then,

$$D_b = 20 \left(\frac{660}{0\cdot 249}\right)^{(1/2\cdot 207)} = 712 \text{ mm}$$

$$S_T = 1\cdot 25\, d_o = 1\cdot 25 \times 20 = 25 \text{ mm}$$

Using equation (5.18) the shell internal diameter will be,

$$D_i = 1\cdot 05 \times 25\ (660 \times 1\cdot 33)^{0\cdot 5} = 778 \text{ mm}$$

Therefore, provide D_s = 780 mm if fabricated from the sheet or suitable standard pipe may be chosen for shell. Choosing a baffle spacing (B_f) as $D_s/4$,

$$B_f = \frac{780}{4} = 195 \qquad \text{say, 200 mm}$$

Using equation (5.35) the cross-flow area is,

$$A_s = \left[\frac{(25 - 20)}{25}\right] \times 0\cdot 78 \times 0\cdot 20 = 0\cdot 0312 \text{ m}^2$$

Then, the linear velocity of steam inside the shell will be,

$$u_s = \frac{G_s}{\rho_s} = \frac{m_s/A_s}{\rho_s} = \frac{21952/3600}{0\cdot 0312 \times 970} = 0\cdot 20 \text{ m/s}$$

An equivalent diameter can be determined by using equation (5.36)

$$d_e = \left(\frac{1\cdot 10}{20}\right)(25^2 - 0\cdot 917 \times 20^2) = 14\cdot 2 \text{ mm}$$

Then, the Reynolds number on shell side will be,

$$N_{Re} = \frac{u_s\, d_e\, \rho_s}{\mu_s} = \frac{(0\cdot 20 \times 3600) \times 0\cdot 0142 \times 970}{0\cdot 97} = 10224$$

For this, the friction factor $f_s = 0.0035$ is obtained from the Fig. 5.8. Then, from equation (5.37), the pressure drop through the shell will be,

$$\Delta p_s = 4 \times 0\cdot 0035\, (0\cdot 20)^2 \times 970 \left(\frac{780}{14\cdot 2}\right)\left(\frac{5\cdot 96}{0\cdot 20}\right)$$

$= 889 \text{ N/m}^2$ *i.e.* $0\cdot 89 \text{ kN/m}^2$

This pressure drop is very small hence, safe

Mechanical design : Assume that the shell is to be fabricated from carbon steel for which the permissible stress is 95 N/mm^2 and that for copper tube is 42 N/mm^2. Shell side pressure is 0·28 N/mm^2 and with 50% additional, the design pressure will be 0·42 N/mm^2. Also, tube side pressure is higher and for copper, the design pressure can be increased by 20% *i.e.* about 3·5 N/mm^2. Let, the nozzle inlet and outlet diameters be 70 mm.

Shell side parameters : Shell thickness is given by equation (3.4) assuming the joint efficiency as $j = 0\cdot 85$, then,

$$t_s = \frac{0\cdot 42 \times 780}{(2 \times 95 \times 0\cdot 85) - 0\cdot 42} = 2\cdot 03 \text{ mm}$$

but, minimum thickness should be 3 mm from Table 3.2 for vessel diameter up to 1000 mm. Hence, provide 6 mm shell thickness with corrosion allowance. Using seamless pipe ($j = 1\cdot 0$) for nozzle, the thickness of nozzle wall is,

$$t_n = \frac{0\cdot 42 \times 70}{(2 \times 95 \times 1) - 0\cdot 42} = 0\cdot 15 \text{ mm}$$

Provide minimum 4 mm thick nozzle wall with the corrosion allowance.

In order to compute the head thickness, equation (3.19) gives the stress intensification factor as below for crown radius R_e = 780/2 = 390 mm and knuckle radius R_n = 39 mm (*i.e.* 10%).

$$S_i = 0\cdot 25\left[3 + \left(\frac{390}{39}\right)^{0\cdot 5}\right] = 1\cdot 54$$

and from equation (3.18) head thickness will be,

$$t_h = \frac{0 \cdot 42 \times 780 \times 1 \cdot 54}{2 \times 95 \times 1} = 2 \cdot 65 \text{ mm}$$

Provide similar head wall thickness as that of the shell *i.e.* 6 mm with corrosion allowance.

Baffles (segmental with 30% cut) of 6 mm thickness shall be provided at 780/5 = 156 mm c/c fixed with 6-10 mm diameter tie rods.

Flanged joint between tube sheet and shell : Selecting compressed asbestos filled stainless steel jacket plain sheet from Table 3.4, design seating stress and gasket factor from Table 3.5 are y = 45 MN/m^2 and m = 3·50. Assuming ($D_i = G_i$ = 780 mm) and equation (3.41) yields,

$$\frac{G_o}{G_i} = \left[\frac{45 - 0 \cdot 42 \times 3 \cdot 5}{45 - 0 \cdot 42\,(3 \cdot 5 + 1)}\right]^{0 \cdot 5} = 1 \cdot 005$$

then G_o = 780 × 1·005 = 783·79 mm say, 784 mm

Thus, the gasket width required is 0·5 (784 – 780) = 2 mm only.

However, provide gasket width of 5 mm, then, G_o = 790 mm and from Table 3.6, $b = b_o$ = 0·5 × 5 = 2·5 mm. Then,

$$G = d_o - 2b = 790 - 2 \times 2 \cdot 5 = 785 \text{ mm}$$

Calculation of bolt loads

$$P_p = 0 \cdot 25\,\pi\,(785)^2 \times 0 \cdot 42 = 203272 \text{ N}$$

$$P_g = 2\pi \times 785 \times 2 \cdot 5 \times 3 \cdot 5 \times 0 \cdot 42 = 18126 \text{ N}$$

then P_o = 203272 + 18126 = 221398 N

also P_b = π × 785 × 2·5 × 45 = 277442 N

Considering the larger load, bolting area required is,

$$A_d = \frac{277442}{138} = 2010 \cdot 45 \text{ mm}^2$$

Providing minimum size as M 12 × 1·5 bolts, root mean area is,

$$A_r = 0 \cdot 25\,\pi\,(12 - 2 \times 1 \cdot 5)^2 = 63 \cdot 62 \text{ mm}^2$$

Then, the number of bolts required is,

$$= \frac{2010 \cdot 45}{63 \cdot 62} = 31 \cdot 60 \quad \text{say, 32 bolts}$$

For the selected bolts B_s = 75 mm and r = 20 mm (Table 3.7).

Assuming 6 mm hub thickness, from equation (3.44),

$$D_p = 780 + 2\,(1 \cdot 415 \times 6 + 20) = 837 \text{ mm}$$

From equation (3.43),

$$D_p = \frac{32 \times 75}{\pi} = 764 \text{ mm}$$

Since, this is less than 780 mm, provide 36 bolts at $B_s = 73$ mm. Then,

$$D_p = \frac{36 \times 73}{\pi} = 836 \cdot 5 \text{ mm}$$

Hence, provide 36-12 mm diameter bolts of 5% CrMo steel (Table 3.8) at 836·5 mm pitch circle diameter, $f_a = 138$ N/mm^2.

Now $$\frac{A_b f_a}{\pi G N} = \frac{36 \times 63 \cdot 62 \times 138}{\pi \times 785 \times 5} = 25 \cdot 63 \text{ N/mm}^2$$

which is $< 2y = 2 \times 45 = 90$ N/mm^2, hence, gasket is safe.

Flange thickness : In equation (3.50),

$$h_G = 0 \cdot 5\ (837 - 785) = 26 \text{ mm}, W_m = 277442 \text{ N}, W_p = 203272 \text{ N}$$

and $$K = \left[0 \cdot 3 + \frac{1 \cdot 5 \times 277442\ (26)}{785 \times 203272}\right]^{-1} = 2 \cdot 72$$

then $$t_f = 785 \left[\frac{0 \cdot 42}{(2 \cdot 72 \times 95)}\right]^{0 \cdot 5} = 31 \cdot 65 \text{ mm}$$

Provide 35 mm thickness for flange considering 3 mm corrosion allowance on each side of wall.

Tube side parameters : Equation (3.6) gives the tube wall thickness as,

$$t_t = \frac{3 \cdot 5 \times 20}{(2 \times 42 \times 1) + 3 \cdot 5} = 0 \cdot 80 \text{ mm}$$

but, provided tube wall thickness is 1.6 mm, which is alright.

Using equation (5.20), the tube sheet thickness will be,

$$t_{ts} = 1 \cdot 25 \times 785 \left[\frac{0 \cdot 25 \times 3 \cdot 5}{95}\right]^{0 \cdot 5} = 94 \cdot 17 \text{ mm}$$

Provide 100 mm thick tube sheets including corrosion allowance.

Channel wall thickness : In equation (5.24), $K = 0 \cdot 3$ for ring type gasket, $G_c = 785$ mm

then $$t_{cc} = 785 \left[\frac{0 \cdot 3 \times 3 \cdot 5}{95}\right]^{0 \cdot 5} = 82 \cdot 53 \text{ mm}$$

Provide 90 mm thickness considering corrosion allowance.

Flanged joint between tube sheet and channel : For higher pressure in tubes and channel, select asbestos filled flat stainless steel jacketed gasket (Tables 3.4 and 3.5) for which $y = 62 \cdot 50$ and $m = 3 \cdot 75$; the dimensions being similar to flanged joint at shell side,

$$\frac{G_o}{G_i} = \left[\frac{62\cdot5 - 3\cdot5 \times 3\cdot75}{62\cdot5 - 3\cdot5\,(3\cdot75 + 1)}\right]^{0\cdot5} = 1\cdot037$$

Then $G_o = 780 \times 1.037 = 808{\cdot}86$ mm say, 810 mm and the gasket width required is $0{\cdot}5\,(810 - 780) = 15$ mm.

From Table 3.6 $b_o = 0{\cdot}5 \times 15 = 7{\cdot}5$ mm and $b = 2{\cdot}5\,(7{\cdot}5)^{0{\cdot}5} = 6{\cdot}84$ mm say, 7 mm and $G = 810 - 2 \times 7 = 796$ mm

Bolt loading will be as under

$$P_p = 0{\cdot}25\,\pi\,(796)^2 \times 3{\cdot}5 = 1{\cdot}742(10)^6 \text{ N}$$

$$P_g = 2\,\pi \times 796 \times 7 \times 3{\cdot}75 \times 3{\cdot}5 = 0{\cdot}46(10)^6 \text{ N}$$

$$P_o = (1{\cdot}742 + 0{\cdot}46)\,(10)^6 = 2{\cdot}202(10)^6 \text{ N}$$

$$P_b = \pi \times 796 \times 7 \times 62{\cdot}5 = 1{\cdot}094(10)^6 \text{ N}$$

Hence, design load is $2{\cdot}202(10)^6$ N and bolting area required is,

$$A_d = \frac{2\cdot202\,(10)^6}{165\cdot8} = 13281 \text{ mm}^2$$

Selecting M 24 × 2 bolts from Table 3.7, $B_s = 60$ to 75 mm and $r = 35$ mm

$$A_r = 0{\cdot}25\,\pi\,(24 - 2 \times 2)^2 = 314{\cdot}16 \text{ mm}^2$$

$$\text{Number of bolts required} = \frac{13281}{314\cdot16} = 42{\cdot}26 \text{ i.e. 44 bolts}$$

With 10 mm hub thickness,

$$D_p = 780 + 2\,(1{\cdot}415 \times 10 + 35) = 878{\cdot}3 \text{ mm}$$

then $$B_f = \frac{878\cdot3\,\pi}{44} = 62{\cdot}71 \text{ mm}$$

Providing $B_s = 63$ mm, $D_p = \dfrac{44 \times 63}{\pi} = 882{\cdot}35$ mm

The difference between the two values is very small and hence, provide larger value of $D_p = 882$ mm.

Also $$\frac{A_b f_a}{\pi\,GN} = \frac{44 \times 314\cdot16 \times 165\cdot8}{\pi \times 796 \times 15} = 61{\cdot}09 \text{ N/mm}^2$$

but, $2y = 2 \times 62{\cdot}5 = 123$ N/mm², hence, provided gasket is safe.

Thickness of nozzle : From equation (3.4),

$$t_n = \frac{3\cdot5 \times 70}{(2 \times 95 \times 0\cdot85) - 3\cdot5} = 1.55 \text{ mm}$$

With 3 mm corrosion allowance provide nozzle wall thickness as 6 mm.

Since, the nozzle is working at higher pressure and has large size, a reinforcing pad is to be provided on the channel cover, the area of compensation being,

$A = d_i t_c = 70 \times 90 = 6300 \text{ mm}^2$

Additional thickness of nozzle and channel cover provide compensation.

$h_o = h_i = 2{\cdot}5\, t_n = 2{\cdot}5 \times 6 = 15 \text{ mm}, C = 3 \text{ mm}$

$D_o = 780 + 2 \times 6 = 792 \text{ mm and } D_n = 70 + 2 \times 6 = 82 \text{ mm}$

then $$t_s' = \frac{3{\cdot}5 \times 792}{(2 \times 95 \times 0{\cdot}85) + 3{\cdot}5} = 16{\cdot}8 \text{ mm}$$

and $$t_n' = \frac{3{\cdot}5 \times 82}{(2 \times 95 \times 0{\cdot}85) + 3{\cdot}5} = 1{\cdot}74 \text{ mm}$$

$A_s = 70\,(90 - 16{\cdot}8 - 3) = 4914 \text{ mm}^2$

$A_o = 2 \times 15\,(6 - 1{\cdot}74 - 3) = 38 \text{ mm}^2$

$A_i = 2 \times 15\,(6 - 2 \times 3) = 0$

Thus, total area available for compensation is,

$A' = 4914 + 38 = 4952 \text{ mm}^2$

Area of compensation plate required is,

$A_p = 6300 - 4952 = 1348 \text{ mm}^2$

then the plate thickness is,

$t_p\,(70 + 6 \times 3 - 2 \times 1.74) = 1348$

and $t_p = 15{\cdot}95 \text{ mm}$

Hence, provide standard 18 mm thick plate for compensation.

5.7 Plate heat exchanger

This type of heat exchanger is popular in dairy and food process industries because of its extremely high rates of heat transfer even at low fluid rates and low temperature gradients, compactness, flexibility in design, low liquid retention and easy dismantling for cleaning. It is constructed by a series of rectangular plates fixed between the two head frames. Each plate has a port at the corner sealed by a gasket and they are connected through the troughs so that the product flows between the two heated surfaces in a very thin film (Fig. 5.18), the pitch of the plates being 1 to 2 or upto 5 mm. Plates are available with an effective area from 0·03 to 1·30 m^2 and up to 400 plates can be accommodated in a large frame. Since, the plates are mounted with the help of bolts, easy dismantling for cleaning is possible. This has also the advantage of increasing or decreasing the heating surface by adding or removing the plates. It is also possible to pass two or three fluids at a time for heating or cooling in plate exchanger. The fluid may also be cooled in one section and heated in another section of the same exchanger.

At low and moderate pressure (< 2·0 N/mm^2) the plate type heat exchangers are competitive with shell and tube exchangers for heat transfer between fluids, especially where, corrosion-resistant materials are required and hence, the units are more compact as they induce turbulence even at low fluid flow rates. The counter current flow of fluids moving through the adjacent plates gives the maximum heat transfer rate. Number of passes can be varied by omitting few ports of the plates. The high heat transfer coefficients (of the order 2400 – 6000 Wm^2K) enable these exchangers to be operated with very low temperature difference, so that a high degree of heat recovery is obtained. The operating temperature is limited by the type of gasket material to about 250 °C. These exchangers are relatively effective with viscous fluids *i.e.* viscosity up to about 300 *P* and hence, they are particularly successful in dairying and brewing industries, where the low liquid capacity and the close control of temperature are the valuable features.

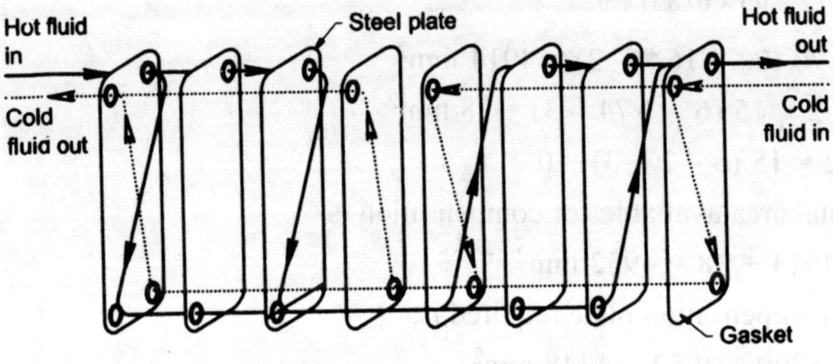

Fig. 5.18 *Plate type heat exchanger*

There are three types of flow patterns in plate heat exchangers namely, series, looped and complex. In series flow pattern, the fluid streams change direction after every vertical pass. In looped flow pattern, the fluid streams divide into parallel flow channels and reconverge again at exit in a single stream. Complex flow pattern is a combination of series and looped flow patterns. The following assumptions are made while designing the plate heat exchanger.

(*a*) An overall heat transfer coefficient is constant.

(*b*) Temperature gradient and flow gradients are negligible.

(*c*) Heat losses to the surrounding are negligible.

(*d*) No phase change of fluid occurs during the flow.

Design considerations : In most of the cases, the developments for predicting the convective heat transfer coefficient in a plate heat exchanger have been based on the heat transfer characteristics from the heating medium to the product across the heating surfaces. The following steps are considered while designing the plate heat exchangers.

(1) Determine the quantity of heat to be transferred by a given hot fluid and log mean temperature difference (θ_{lm}) using equation (5.1).

(2) Compute the heat transfer area required using the equation,

$$A_T = \frac{q}{U_o\, F\, \theta_{lm}} \tag{5.38}$$

Then decide the number of plates required considering the effective area of plate.

(3) Determine the values of Reynolds number and Prandlt number given by the relationship as below.

$$N_{Re} = \frac{G_f b}{\mu} \tag{5.39}$$

and $$N_{Pr} = \frac{C_p \mu}{k} \tag{5.40}$$

(4) An average heat transfer coefficients for both hot and cold sides can be determined by using the following equations depending on the type of flow *i.e.* turbulent or laminar, respectively.

$$N_{Nu} = \frac{hb}{k} = 0{\cdot}26\, N_{Re}^{\,0{\cdot}65}\, N_{Pr}^{\,0{\cdot}4} \left(\frac{\mu}{\mu_w}\right)^{0{\cdot}14} \tag{5.41}$$

and $$N_{Nu} = C_1 \left[\frac{N_{Re} N_{Pr}\, d}{L}\right]^{0{\cdot}33} \left(\frac{\mu}{\mu_w}\right)^{0{\cdot}4} \tag{5.42}$$

The coefficient C is depending on the plate geometry, taken as 1·86 to 4·50 the tube wall temperature is determined by the following relationship,

$$T_w = \left[\left(\frac{T_{hi} + T_{ho}}{2}\right) + \left(\frac{h_{ci}}{h_h}\right)\left(\frac{T_{ci} + T_{co}}{2}\right)\right]\left[1 + \frac{h_c}{h_h}\right]^{-1} \tag{5.43}$$

Determine the heat transfer coefficients assuming (μ/μ_w) as unity in equation (5.41) and (5.42) as the case may be and then convert these coefficients using values of respective viscosity at plate wall temperature.

(5) Compute the overall heat transfer coefficient using the following equation.

$$\frac{1}{U_o} = \frac{1}{h_h} + \frac{1}{h_c} + \frac{x_w}{k} + r_h + r_c \tag{5.44}$$

(6) Determine the heat transfer area required in this case and compare with earlier determined using equation (5.38). The later should be at least equal to or higher than this area so that the design is alright.

(7) Pressure drop of flow through plates can be determined using the following equation.

$$\Delta p = \left(\frac{8 f\, L_n}{b}\right)\left(\frac{G_f^{\,2}}{2\rho}\right) \tag{5.45}$$

where, f is given approximately by the relationship as below,

$$f = 1{\cdot}25\, N_{Re}^{\,-0{\cdot}3} \tag{5.46}$$

The mean temperature difference is generally higher in plate exchangers than in shell and tube exchangers, as the flow arrangement gives a closer approach to true counter current flow. For a series of arrangements, the correction factor for θ_{lm} may be close to 1·0. For normal design conditions, the fluid flow is laminar when $N_{Re} < 800$ and turbulent when $N_{Re} > 800$. However, most of the plate heat exchangers, have turbulent flow conditions even at low flow rates. The flow will turn from laminar to turbulent at Reynolds number of 100 to 400, depending on the design of the plates used. With some designs, the turbulence can be achieved at very low Reynolds number, which makes the viscous fluids suitable for plate heat exchangers.

The same pressure drop for counter current flow of fluids can be obtained by providing thick gaskets on the heating medium side which, in turn, increase the gap between plates resulting in a decrease of medium flow. Thus, considering the product/medium ratio (higher for higher values of θ_{lm}), the required velocities of the fluids can be determined.

Example 5.7 : *Viscous liquid is to be cooled from 95 to 60 °C in a looped flow steel plate heat exchanger flowing at the rate of 2·0 kg/s. The cooling medium is water available at 18 °C leaving the exchanger at 44 °C. The inlet and outlet ports are 800 mm apart from each other. One millimetre thick plates have effective width of 600 mm, the projected heat transfer area being 2·75 $(10)^5$ mm² fitted with 3 mm thick gaskets. The viscosity of hot fluid and water is 0·122 and 7·8 $(10)^{-4}$ Ns/m², heat capacity 3200 and 4310 J/kg °C, thermal conductivity 0·195 and 0·82 W/m °C, respectively. At the plate wall temperature, viscosity of respective fluids may be assumed as 0·112 and 5·9$(10)^{-4}$ Ns/m², correction factor as 0·93, overall heat transfer coefficient as 190 W/m²K and thermal conductivity for steel plates as 19·2 WmK. Check the exchanger parameters and pressure drop.*

Solution. Total heat load from hot viscous fluid is,

$$q = m_h \, C_{ph} \, \Delta T_h = 2{\cdot}0 \times 3200 \, (95 - 60) = 224000 \text{ W}$$

Assuming no heat loss, this heat is transferred to water. Then, the water flow rate is,

$$m_c = \frac{224000}{[4310\,(44-18)]} = 2.0 \text{ kg/s}$$

and from equation (5.1),

$$\theta_{lm} = \frac{(95-44)-(60-18)}{\ln[(95-44)/(60-18)]} = 46{\cdot}35 \text{ °C}$$

Total heat transfer area of plate required is,

$$A_T = \frac{q}{U_o F \theta_{lm}} = \frac{224000}{190 \times 0{\cdot}93 \times 46{\cdot}35} = 27{\cdot}35 \text{ m}^2$$

Then number of plates required is,

$$N = \frac{27 \cdot 35}{[2 \cdot 75\,(10)^5 \times 10^{-6}]} = 99{\cdot}45$$

Hence, provide 100 plates. For looped flow arrangement, the number of streams of hot fluid is,

$$n_h = \frac{(100+1)}{2} = 50 \cdot 5 \text{ say } 51$$

and for cold water, $n_c = n_h + 4 = 55$

Hot fluid flow per stream is,

$$m_h = \frac{2 \cdot 0}{51} = 0{\cdot}039 \text{ kg/s}$$

and the flow per unit area will be,

$$G_h = \frac{m_h}{A_f} = \frac{0 \cdot 039}{[600 \times 3 \times 10^{-6}]} = 21{\cdot}67 \text{ kg/sm}^2$$

Then $$N_{Re} = \frac{G_h b}{\mu} = \frac{21 \cdot 67 \times 3 \times 2\,(10)^{-3}}{0 \cdot 122} = 1 \cdot 06$$

and $$N_{Pr} = \frac{C_p\,\mu}{k} = \frac{3200 \times 0 \cdot 122}{0 \cdot 195} = 2002$$

For cold fluid (water), mass flow rate per stream is,

$$m_c = \frac{2 \cdot 0}{55} = 0{\cdot}036 \text{ kg/s}$$

Then $$G_c = \frac{0 \cdot 036}{600 \times 3 \times 10^{-6}} = 20{\cdot}0 \text{ kg/sm}^2$$

also $$N_{Re} = \frac{20 \cdot 0 \times 3\,(10)^{-3} \times 2}{[7 \cdot 8\,(10)^{-4}]} = 153 \cdot 85$$

and $$N_{Pr} = \frac{4310 \times 7 \cdot 8\,(10)^{-4}}{0 \cdot 82} = 4 \cdot 10$$

Assuming (μ/μ_w) as unity and $C = 4{\cdot}5$, for laminar flow, the Nusselt number using equation (5·41) is determined as below.

$$N_{Nu} = 4.5\ \{1{\cdot}06 \times 2002 \times [3\ (10)^{-3}/0{\cdot}8]\}^{0{\cdot}33} = 8{\cdot}92$$

then $$h_h = \frac{N_{Nu} k}{b} = \frac{8 \cdot 92 \times 0 \cdot 195}{[2 \times 3\,(10)^{-3}]} = 289{\cdot}90 \text{ W/m}^2\ {}^\circ\text{C}$$

For cold fluid,

$$N_{Nu} = 4\cdot5\left\{153\cdot85 \times 4\cdot10 \times \left[\frac{3\,(10)^{-3}}{0\cdot8}\right]\right\}^{0\cdot33} = 5\cdot98$$

and $$h_c = \frac{N_{Nu}\,k}{b} = \frac{5\cdot98 \times 0\cdot82}{2 \times 3\,(10)^{-3}} = 817\cdot27 \text{ W/m}^2\,°\text{C}$$

The plate wall temperature can be determined using equation (5.46)

$$T_w = \left[\left(\frac{95+60}{2}\right) + \left(\frac{817\cdot27}{289\cdot90}\right)\left(\frac{44+18}{2}\right)\right]\left[1 + \left(\frac{817\cdot27}{289\cdot90}\right)\right]^{-1}$$
$$= 43\cdot17\,°\text{C}$$

Thus, the heat transfer coefficients on hot and cold sides at 45.36 °C plate wall temperature will be,

$$h_h = 289\cdot90\left(\frac{0\cdot122}{0\cdot112}\right)^{0\cdot4} = 300 \text{ W/m}^2\,°\text{C}$$

and $$h_c = 817\cdot27\left[\frac{7\cdot8\,(10)^{-4}}{5\cdot9\,(10)^{-4}}\right]^{0\cdot4} = 913\cdot82 \text{ W/m}^2\,°\text{C}$$

Assuming fouling factor on hot and cold sides as 0.0002 each an overall heat transfer coefficient using equation (5.44) can be determined as below.

$$\frac{1}{U_o} = \frac{1}{300} + \frac{1}{913\cdot82} + \frac{1(10)^{-3}}{19\cdot2} + 0\cdot0002 + 0\cdot0002$$
$$= 4\cdot89\,(10)^{-3}$$

or $$U_o = 204\cdot93 \text{ W/m}^2\,°\text{C}$$

Then, the heat transfer area required will be,

$$A_T = \frac{q}{U_o F \theta_{lm}} = \frac{224000}{204\cdot93 \times 0\cdot93 \times 46\cdot36} = 25\cdot37 \text{ m}^2$$

but, A_T provided is 27·35 m² hence, the design is alright.

The friction factor for hot and cold sides using equation (5.46) will be,

$$f_h = 1\cdot25\,(1\cdot06)^{-0\cdot3} = 1\cdot228$$

and $$f_c = 1\cdot25\,(153\cdot85)^{-0\cdot3} = 0\cdot276$$

Then, the respective pressure drop using equation (5.45) will be,

$$\Delta p_h = \frac{8 \times 1\cdot228 \times (51 \times 0\cdot8)\,(21\cdot67)^2}{[2 \times 3\,(10)^{-3} \times 960] \times 2} = 16338 \text{ N/m}^2 = 16\cdot34 \text{ kN/m}^2$$

and $$\Delta p_c = \frac{8 \times 0\cdot276 \times (55 \times 0\cdot8)\,(20)^2}{[2 \times 3\,(10)^{-3} \times 996] \times 2} = 3251 \text{ N/m}^2 = 3\cdot25 \text{ kN/m}^2$$

Since, the pressure drop on both sides is less, the design is OK.

Nomenclature

A : cross section area of annulus in double pipe heat exchanger, m^2
A_c : cross section area of annulus on cold side, m^2
A_h : cross section area of annulus on hot side, m^2
A_o : total outer surface of tubes, m^2
A_s : cross section area of shell, m^2
A_T : required heat transfer area of plates, m^2
A_t : cross section area of tubes, m^2
a : area occupied by each tube, m^2
a_n : total area occupied by tubes, m^2
B_f : baffle spacing, m
b : distance between two parallel heated plates, mm
C : corrosion allowance, mm
C_p : heat capacity at constant pressure, kJ/kg °C
C_{pc} : specific heat of cold fluid, kJ/kg °C
C_{ph} : specific heat of hot fluid, kJ/kg °C
C_{ps} : specific heat of fluid on shell side, kJ/kg °C
C_{pt} : specific heat of fluid on tube side, kJ/kg °C
C_1 : coefficient depending on plate geometry.
D_b : tube bundle diameter, m
D_c : thermal diffusivity of fluid on cold side, m^2/s
D_e : equivalent diameter of shell, m
D_h : thermal diffusivity of fluid on hot side, m^2/s
D_i : inside diameter of shell, m
D_o : outer diameter of shell, m
D_p : pitch circle diameter, m
D_n : outer diameter of nozzle, m
d : distance between two parallel plates, mm
d_i : inner diameter of tube, m
d_e : equivalent or hydraulic diameter of annulus, m
d_o : outer diameter of tube, m
d_t : diameter of tube, m
d_1 : outer diameter of inner tube, m
d_2 : inner diameter of outer tube, m
E_s : modulus of elasticity of shell wall material, N/mm^2
E_t : modulus of elasticity of tube wall material, N/mm^2

F : correction factor for heat transfer in plate heat exchanger
f : friction factor in plate heat exchanger
f' : friction factor in double pipe heat exchanger
f_a : permissible stress in material at design temperature, N/mm^2
f_{as} : permissible stress in shell wall material, N/mm^2
f_{at} : permissible stress in tube wall material, N/mm^2
f_c : friction factor on cold side in plate heat exchanger
f_h : friction factor on hot side in plate heat exchanger
f_s : friction factor on shell side
f_t : friction factor on tube side
G : gasket mean diameter for shell, mm
G' : gasket mean diameter for channel cover, mm
G_c : mass flow rate of fluid per unit area on cold side, kg/h m^2
G_f : mass flow rate of fluid per unit area, kg/h m^2
G_h : mass flow rate of fluid per unit area on hot side, kg/h m^2
G_i : inner diameter of gasket, mm
G_o : outer diameter of gasket, mm
G_s : mass flow rate of fluid per unit area on shell side, kg/h m^2
G_t : mass flow rate of fluid per unit area on tube side, kg/h m^2
g : gravitational constant, m/s^2
h_c : surface heat transfer coefficient for cold fluid, W/m^2 °C
h_h : surface heat transfer coefficient for hot fluid, W/m^2 °C
h_i : surface heat transfer coefficient for inside surface, W/m^2 °C
h_o : surface heat transfer coefficient for outside surface, W/m^2 °C
h_{sc} : scale heat transfer coefficient for cold side surface, W/m^2 °C
h_{sh} : scale heat transfer coefficient for hot side surface, W/m^2 °C
h_{si} : scale heat transfer coefficient for inside surface of inner pipe, W/m^2 °C
h_{so} : scale heat transfer coefficient for outside surface of inner pipe, W/m^2 °C
j : joint efficiency factor
K : coefficient depending on material and geometry of shell and tube
K_b : multiplying factor for increased area due to baffles
K_s : multiplying factor for tube sheet thickness
k : thermal conductivity of material, W/m °C
k_c : thermal conductivity of cold fluid, W/m °C
k_f : thermal conductivity of fluid, W/m °C
k_h : thermal conductivity of hot fluid, W/m °C

k_t	: thermal conductivity of fluid on tube side, W/m °C
k_l	: coefficient depending on number and pitch of tubes
L	: length of vertical plane, m
L_e	: effective length of tube, m
L_n	: total length of flow path, m
L_s	: length of shell, m
L_t	: length of tubes, m
L_v	: Latent heat of vaporization, kg/kg
m_c	: mass flow rate of cold side fluid, kg/h
m_h	: mass flow rate of hot side fluid, kg/h
m_s	: mass flow rate of shell side fluid, kg/h
m_t	: mass flow rate of tube side fluid, kg/h
N	: number of plates required in heat exchanger
N_{Nu}	: Nusselt number
N_{Pr}	: Prandtl number
N_{Prc}	: Prandtl number for cold side fluid
N_{Prh}	: Prandtl number for hot side fluid
N_{Re}	: Reynolds number
N_{Rec}	: Reynolds number for cold side fluid
N_{Reh}	: Reynolds number for hot side fluid
N_{St}	: Stanton number
N_t	: number of passes of fluid through heat exchanger
n	: total number of tubes
n_h	: number of streams of hot fluid in plate heat exchanger
n_p	: number of tubes per pass
n_l	: exponent depending on number and pitch of tubes
P	: perimeter of inner tube outer surface in double pipe heat exchanger, m
P_s	: load exerted on shell wall, N
P_t	: load exerted on tube wall, N
p	: design pressure, N/mm^2
p_s	: stress due to shell wall load, N/mm^2
p_t	: stress due to tube wall load, N/mm^2
R	: capacity ratio in heat exchanger
r_c	: fouling factor on cold side
r_h	: fouling factor on hot side
r_i	: inner radius of tube, mm

r_o : outer radius of tube, mm
S_T : pitch of the tubes, mm
T_{ci} : temperature of cold fluid at inlet, °C
T_{co} : temperature of cold fluid at outlet, °C
T_{hi} : temperature of hot fluid at inlet, °C
T_{ho} : temperature of hot fluid at outlet, °C
T_j : bulk (mean) temperature on jacket side fluid, °C
T_p : bulk (mean) temperature on pipe side fluid, °C
T_{sat} : vapour temperature at saturation point, °C
T_{si} : temperature of shell side fluid at inlet, °C
T_{so} : temperature of shell side fluid at outlet, °C
T_{ti} : temperature of tube side fluid at inlet, °C
T_{to} : temperature of tube side fluid at outlet, °C
T_w : estimated temperature of tube wall, °C
t_{cc} : thickness of channel cover wall, mm
t_s : thickness of calendria shell, mm
t_t : tube wall thickness, mm
t_{ts} : tube sheet thickness, mm
U_o : overall heat transfer coefficient, W/m^2 °C
u_c : velocity of fluid on cold side, m/s
u_h : velocity of fluid on hot side, m/s
u_s : velocity of fluid on shell side, m/s
u_t : velocity of fluid on tube side, m/s
V_t : volumetric flow of fluid through tubes, m^3/s
w : specific weight of fluid, MN/m^3
x_w : thickness of wall plate, mm
θ : temperature ratio
θ_{am} : arithmetic mean temperature difference, °C
θ_i : hot and cold fluids temperature difference at inlet, °C
θ_{lm} : log mean temperature difference, °C
θ_o : hot and cold fluids temperature difference at outlet, °C
δ_L : mean thermal elongation of length, m/m
δ_{Ls} : mean thermal elongation of shell length, m/m
δ_{Lt} : mean thermal elongation of tube length, m/m
Δp : pressure drop through heat exchanger, N/m^2
Δp_c : pressure drop through heat exchanger on cold side, N/m^2

Δp_h : pressure drop through heat exchanger on hot side, N/m^2

Δp_s : pressure drop through heat exchanger on shell side, N/m^2

Δp_t : pressure drop through heat exchanger on tube side, N/m^2

Δp_{tm} : maximum permissible pressure drop through tubes, N/m^2

Δt : hot and cold fluids temperature difference at inlet, °C

Δt_h : temperature reduction in hot fluid, °C

ρ : density of fluid, kg/m^3

ρ_c : density of fluid on cold side, kg/m^3

ρ_h : density of fluid on hot side, kg/m^3

ρ_s : density of fluid on shell side, kg/m^3

ρ_t : density of fluid on tube side, kg/m^3

μ : viscosity of fluid at bulk fluid temperature, kg/h m

μ_s : viscosity of fluid on shell side, kg/h m

μ_t : viscosity of fluid on tube side, kg/h m

μ_w : viscosity of fluid at tube wall temperature, kg/h m

υ : kinetic viscosity of fluid, kg/h m

6

Evaporators

6.1 Introduction

Evaporators are employed for concentration of the fluid food such as fruit juices to reduce the volume of product without the loss of nutrients to a significant quantity. The reduction in the quantity enables efficient storage and transportation of the product. It also helps to feed a product at relatively low moisture content and in reduced quantity after evaporation. Widely varying characteristics of the product have the specific influence on the design of evaporator. An evaporator raises the product temperature to its boiling point and holding for the time required to reach the desired concentration by evaporating free moisture present in the product. Use of low pressure or high vacuum makes it possible to reduce substantial quantity of water without any qualitative loss of the heat-sensitive products.

6.2 Construction

An evaporator consists of essential four parts, namely (*a*) an evaporation chamber, (*b*) a source of heating, (*c*) facility for condensing hot fluid or condenser and (*d*) a propeller for removal of the non-condensing gases. In majority of evaporators the heat transfer is achieved through tubular surfaces, enclosed in a cylindrical shell. The liquid to be concentrated flows through tubes and heating fluid, usually low pressure, dry, saturated steam flows over the tubes vaporizing the liquid. Due formation of vapours, natural circulation is created or for more viscous liquids forced circulation is provided either by a vacuum pump or a propeller. A foam is formed due to boiling of liquid and hence, the vapour contains liquid droplets suspended in it. Vapour release chamber is provided for their separation at the top of tubes in evaporation chamber. Baffles and impingement, centrifugal separator or mesh pod can be used for the separation. A condenser is necessary for the condensation of vapour produced during the operation.

6.3 Types of evaporators

Evaporators are made of vertical, horizontal or inclined tubes, may or may not be submerged in the solution in vessel such as oblong or helical pipe coils. Depending on tube arrangement, method of circulation, heating surface and heating medium, the evaporators are classified in five groups by Moore and Hessler (1963) described as below.

(a) Short vertical tube evaporator (calendria)

It consists of a vertical cylindrical shell, (Fig. 6.1) 1 to 6 m diameter, incorporated with a bundle of short vertical tubes called as the calendria (25 to 75 mm outside diameter) and 750 to 2000 mm height, usually 150 mm fixed in two horizontal tube sheets bolted to shell with the help of flanges, located at the bottom of the shell which may be conical or dished. The flanged corrugated sections are bolted together to form a vapour chamber above the calendria for sufficient height. The product or solution passes vertically through the tubes and moves down by natural circulation through the central downtake which has the area equal to 40 to 100% of the total cross sectional area of tubes. If higher rates of circulation are desired, a propeller can also be used. In order to reduce the scaling inside the tubes, multiple downtake should be provided. The steam flow-outside the tubes for heating uniformly through the baffles in the steam chest of the evaporator. The non-condensable gas (air) is taken out from the top of the vapour chamber whereas the condensate is withdrawn from a convenient point near the lower tube sheet. For the steam inlet, condensate outlet, solution inlet/outlet, vapour outlet, manhole, sight glass, level gauge etc. the nozzles are to be provided at convenient locations. These nozzles and the supporting legs are welded to the shell body. This evaporator is designed as a shell with internal steam pressure with tube sheets of thickness based on equation (5.20) and Table 5.6.

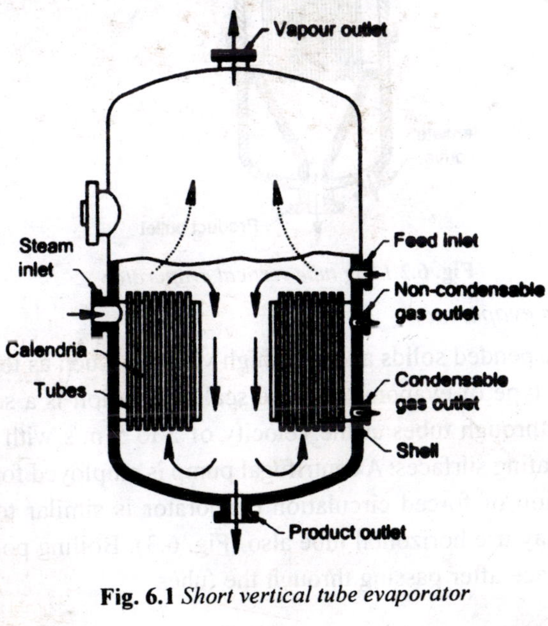

Fig. 6.1 *Short vertical tube evaporator*

(*b*) *Long tube vertical evaporator*

This evaporator is provided with relatively long tubes (3 to 6 m) of 25 to 50 mm diameter, enclosed inside the shell and projecting into the vapour chamber as shown in Fig. 6.2. A baffle is provided over the open ends of tubes at upper tube sheet to facilitate natural circulation of unvaporised liquid through return pipe. The liquor feed enters the bottom of the tubes, heated by steam surrounded around them and starts boiling while moving up the tubes. A mixture of vapour and liquor impinges on the baffle at high velocity which breaks the foam and primary separation takes place. This type of evaporator is limited to products with low to moderate viscosity and are foamy or frothy in nature.

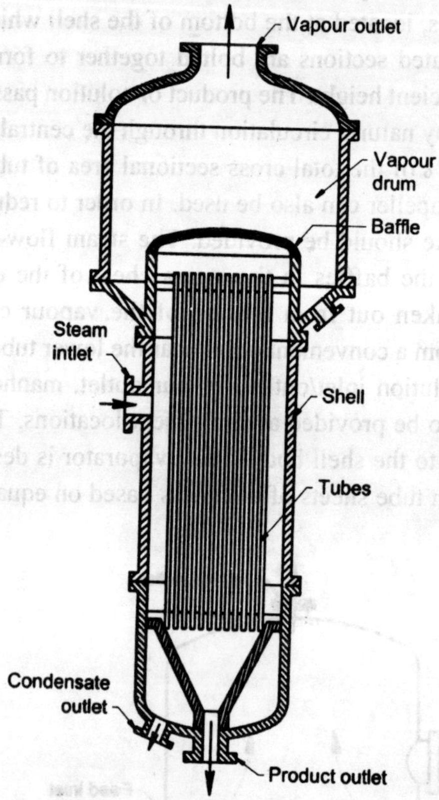

Fig. 6.2 *Long tube vertical evaporator*

(*c*) *Forced circulation evaporator*

Products with suspended solids and with high viscosity such as tomato products can be evaporated in this type of evaporator where scale formation is a serve problem. The product is circulated through tubes at the velocity of 2 to 5 m/s with high heat transfer rates requiring less heating surfaces. A centrifugal pump is employed for circulation of the liquor. The construction of forced circulation evaporator is similar to the vertical tube evaporators or they may use horizontal tube also (Fig. 6.3). Boiling point elevation takes place in separating space after passing through the tubes.

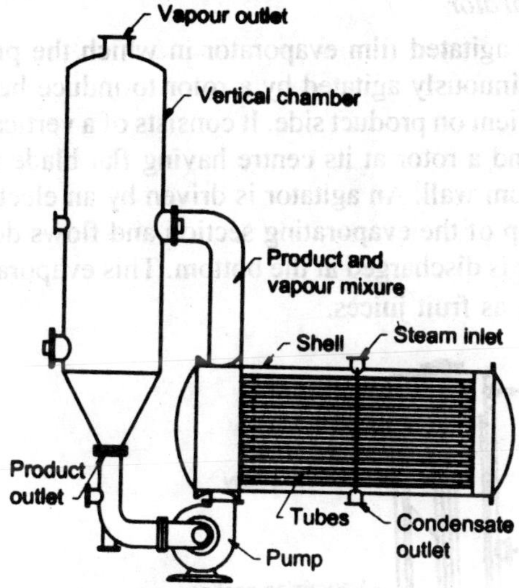

Fig. 6.3 *Horizontal tube forced circulation evaporator*

In a horizontal tube evaporator a shell is either a vertical cylindrical or square with horizontal square tube bundle enclosed at bottom (Fig. 6.4). The steam is circulated through tubes introduced and collected by two separate channels at both the ends of tubes and the product surrounds them. Thick solution (liquor) is removed from the bottom and vapour at the top of the shell. This evaporator uses small tubes (20 to 30 mm outside diameter) with shell diameter of 1 to 4 m and 2·5 m height. It is not suitable for scale forming liquors but used for those products where the final product is liquid.

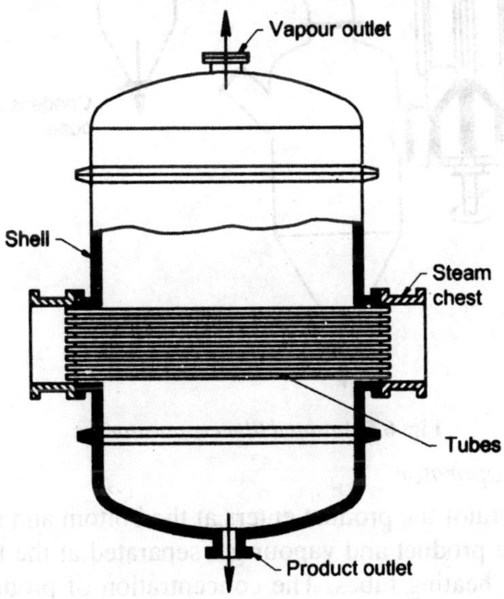

Fig. 6.4 *Horizontal tube evaporator*

(*d*) *Agitated film evaporator*

Fig. 6.5 shows an agitated film evaporator in which the product film near the heat transfer surface is continuously agitated by a rotor to induce heat transfer and enhances the heat transfer coefficient on product side. It consists of a vertical shell (may be jacketed) for steam circulation and a rotor at its centre having flat blade type fins with about one millimetre clearance from wall. An agitator is driven by an electric motor at the top. The product enters at the top of the evaporating section and flows down the walls, vaporised and concentrated liquor is discharged at the bottom. This evaporator is able to handle heat sensitive products such as fruit juices.

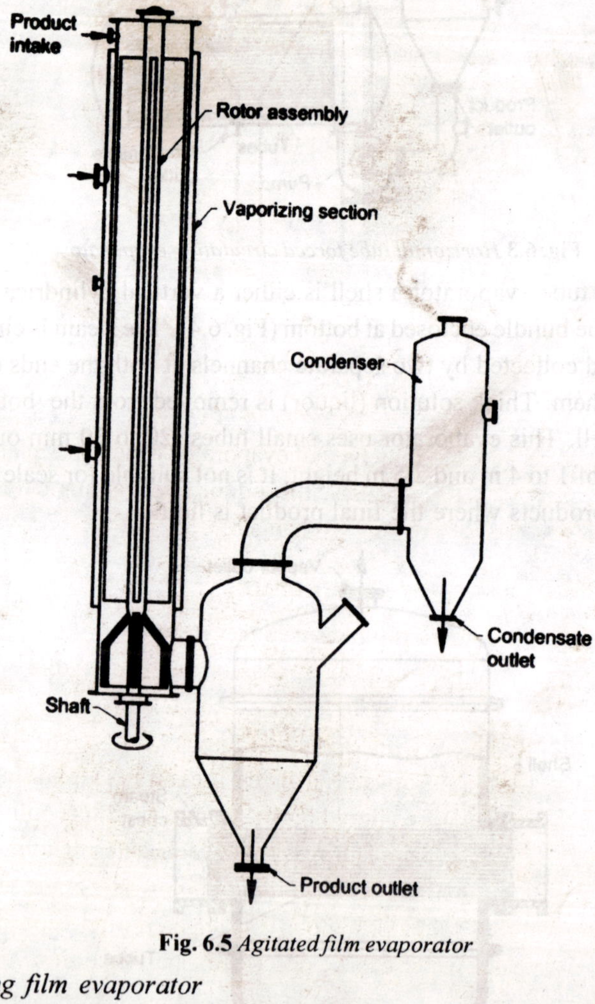

Fig. 6.5 *Agitated film evaporator*

(*e*) *Rising-failing film evaporator*

In this type of evaporator the product enters at the bottom and carried upward in the form of a rising film. The product and vapour get separated at the top of tubes and then moves down through the heating tubes. The concentration of product is achieved more while moving down. Finally, the product is collected at the bottom of the evaporator for exit. (Fig. 6.6).

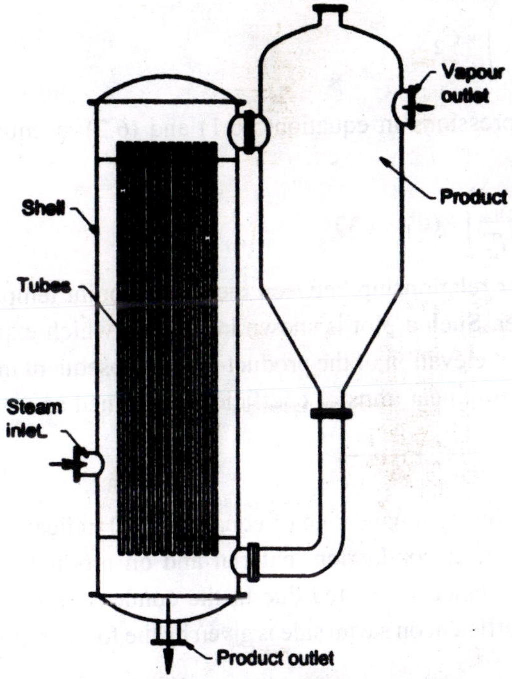

Fig. 6.6 *Rising-falling film evaporator*

A plate evaporator is an additional type of evaporator which uses plate heat exchanger for vaporization of milk and its products, the liquor and vapour being separated in a chamber provided at the end of the heat exchanger.

6.4 Evaluation of evaporator parameters

The boiling point of the liquid products is higher than that of the pure water at a given pressure. This effect is the result of product composition containing solutes having molecular weights that cause the boiling point to be elevated above that of the pure water. Thus, a boiling point elevation (or rise) is defined as the difference between the boiling point of the liquid product and that of the pure water at given pressure. The boiling point elevation magnitude is higher for concentrated liquid products than the dilute solutions. The boiling point elevation can be estimated from During's rule which states that the boiling point of a given liquid is a linear function of the boiling point of pure water at the same pressure. The phase of water from liquid to steam or vapour is changed during vaporization. The latent heat of vaporization for pure water can be described as a function of pressure as below.

$$\ln p = \left(\frac{\lambda_w}{RT_a}\right) + C_1 \tag{6.1}$$

Similar expression can be written for product containing solids and other components which affect heat of vaporization as below.

$$\ln p' = \left(\frac{\lambda'_p}{RT_a'}\right) + C_2 \tag{6.2}$$

Equating the expressions in equations (6.1) and (6.2) at equal vapour pressure is given by,

$$\left(\frac{\lambda'_p}{RT_a'}\right) = \left(\frac{\lambda_w}{RT_a}\right) + (C_1 - C_2) \tag{6.3}$$

which shows the linear relationship between the boiling point temperatures of the liquid product and pure water. Such a plot is shown in Fig. 6.7 which explains that there is an increase in boiling point elevation of the product with a subsequent increase in the product concentration. The overall heat transfer coefficient is defined as.

$$\frac{1}{U_o A_m} = \frac{1}{h_s A_s} + \frac{x}{k A_m} + \frac{1}{h_p A_p} \tag{6.4}$$

Each of the terms on right hand side of equation (6.4) indicate the resistance to heat transfer on steam side, heat conduction material and on product side, respectively. On heating surface the resistance is created due to the condensation film of steam and the surface heat transfer coefficient on steam side is given by the following empirical relationship (Brown *et al.*, 1950).

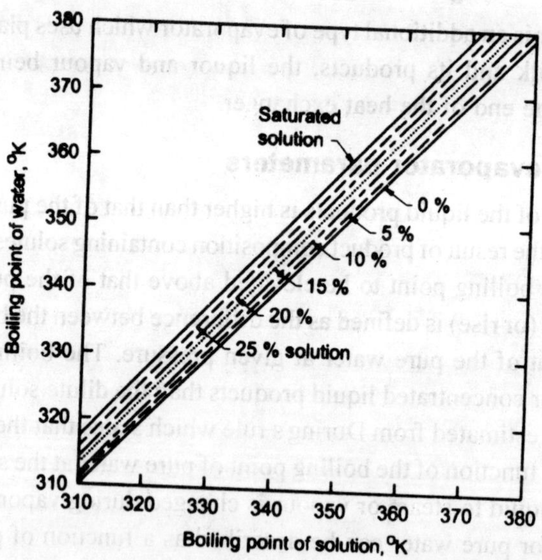

Fig. 6.7 *Boiling point elevation with concentration*

For vertical surface Krieth (1965) presented that the coefficient in equation (6.5) should be 0·94 instead of 1·13. The film coefficient for condensate collecting on horizontal tubes is given by the following equation presented by Charm (1978).

$$h_s = 1{\cdot}18\left[\frac{K_f^3\,\rho_f^2\,g\,\pi\,D_s}{L\,\mu_f\,M}\right]^{1/3} \tag{6.6}$$

An alternative expression was first presented by Bromely (1950) as,

$$h_s = 0{\cdot}725\left[\frac{K_f^3\,\rho_f^2\,g\,\lambda_w}{\mu_f\,D_s(T_s - T_w)}\right]^{1/4} \tag{6.7}$$

in which the quantity of heat transfer involved in phase change is taken into account. The film coefficient for product side is the complex resistance. However, it is determined as described for the tubular heat exchanger *i.e.*, from equations (5.10) and (5.11) under particular flow conditions. Coulson and Richardson (1978) suggested 25% increase in these equations to take care for the additional contribution of boiling heat transfer during convection in tubes.

Capacity of an evaporator is defined as the quantity of water (or solvent) evaporated per unit time. It is proportional to the rate of heat transfer when feed solution is at the boiling temperature corresponding to the pressure in vapour chamber of an evaporator when all the heat is available for vaporization of the solvent. If feed solution is at higher temperature than boiling point, at given pressure, a part of solvent is evaporated adiabatically resulting in higher capacity than an compared with that of the rate of heat transfer. But the capacity is reduced if the feed solution is cold due to additional heat requirement for raising the temperature of feed to its boiling point.

An efficiency of the evaporation process is given by the steam economy which is defined as the quantity of water (or solvent) evaporated from the feed solution per unit quantity of steam utilized, expressed as, $SE = m_v/m_s$.

With an increase in pressure, the latent heat of vaporization for solvent decreases which results in the reduction of steam economy less than unity as in case of single effect evaporators. Higher steam economy can be achieved by utilization of the produced vapour as used in the multiple effect evaporators. Because of use of vapour from one effect as heating source in the next effect, the boiling of feed solution takes place at low temperature and pressure. Steam consumption is the ratio of evaporator capacity of the steam economy.

6.5 Single effect system

When the feed solution passes through one evaporator chamber and heat is also supplied to only one heat transfer area, it is called as a single effect system *e.g.* all the

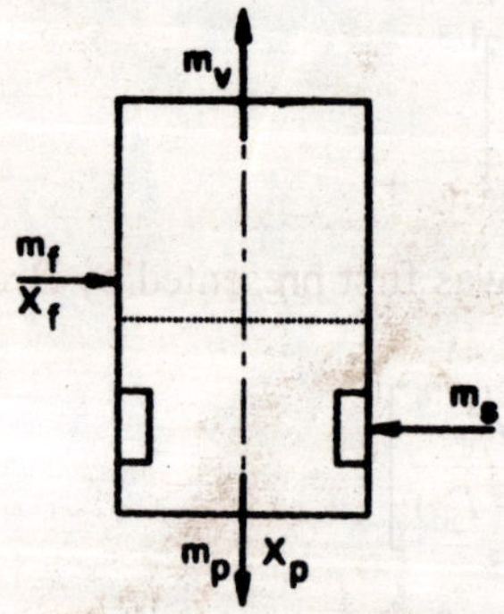

Fig. 6.8 *Mass and enthalpy balance on single effect evaporator*

evaporators described above come under this group. Considering overall mass balance from Fig. 6.8 gives.

$$m_f = m_v + m_p \tag{6.8}$$

and $$m_f X_f = m_p X_p \tag{6.9}$$

Considering a negligible heat loss from the system, an enthalpy balance over the evaporator is that the heat associated with feed and steam is equal to that associated with vapour and product, given by the following equation.

$$m_f H_f + m_s (H_{vs} - H_{cs}) = m_v H_v + m_p H_p \tag{6.10}$$

From equation (6.8),

$$m_v = m_f - m_p \tag{6.11}$$

Putting this value in equation (6.10) and rearranging,

$$m_s \ (H_{vs} - H_{cs}) = (m_f - m_p) H_v + m_p H_p - m_f H_f \tag{6.12}$$

With specific heat of feed solution fed and the temperature in the evaporator to boiling point, equation (6.12) can be written as below.

$$m_s (H_{vs} - H_{cs}) = m_f C_{pf} (T_e - T_f) + (m_f - m_p) \lambda_v \tag{6.13}$$

When the boiling point rise is negligible $\lambda_v = \lambda_w$ which can be directly, read from steam tables. The area of heat transfer for an evaporator can be calculated from the basic formula for rate of heat transfer where the temperature differences of $(T_s - T_c)$ is considered. Equation (6.8) through (6.13) can be used for the design of a single effect evaporator. Based on components of food product Charm (1978) proposed the following equation for determination of specific heat of the product.

$$C_{pp} = C_{ps} X_s + C_{pw} X_w \tag{6.14}$$

6.6 Multiple effect system

The water vapour is released in case of a single effect system to the surrounding atmosphere whereas, it is utilized in the following step to economise the use of steam. The vapours from the first evaporator (or effect) are introduced into the evaporation chamber of the second evaporator while the feed solution loses water flowing in the same direction

(Fig. 6.9). The total concentration desired in achieved in three steps of such a triple effect system. Vapours produced in first two steps are utilized as medium media in the preceding step. Although, there is no theoretical limit to the number of steps or effects, a practical limit is reached when the plant cost exceeds the steam saving. A multiple effect system is also designed writing a mass and enthalpy balance equations for each effect and for the entire system. For forward feed system shown in Fig. 6.9, the mass balance equation is as below.

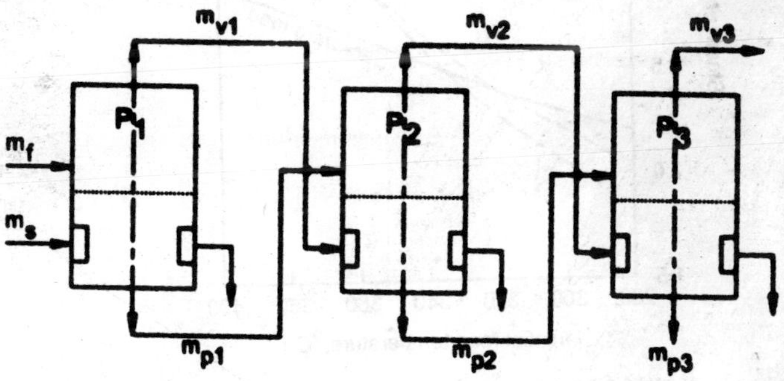

Fig. 6.9 *Mass and enthalpy balance on multiple effect evaporator*

$$m_f = m_{v1} + m_{v2} + m_{v3} + m_p \tag{6.15}$$

and $$m_f X_f = m_p X_p \tag{6.16}$$

which considers the changes in the product and its solid content, respectively. Similarly, the enthalpy balance for the system can be written as below.

$$m_s H_s + m_f H_f = m_{v1} H_{v1} + m_{p1} H_{p1} \tag{6.17}$$

$$m_{p1} H_{p1} + m_{v1} H_{v1} = m_{v2} H_{v2} + m_{p2} H_{p2} \tag{6.18}$$

$$m_{p2} H_{p2} + m_{v2} H_{v2} = m_{v3} H_{v3} + m_{p3} H_{p3} \tag{6.19}$$

The heat transfer area for each effect can be determined by considering the heat transfer equations for all the effect as below.

$$q_1 = U_1 A_1 (T_s - T_1) = m_s (H_{vs} - H_{cs}) \tag{6.20}$$

$$q_2 = U_2 A_2 (T_1 - T_2) = m_{v1} (H_{v1} - H_{c1}) \tag{6.21}$$

$$q_3 = U_2 A_3 (T_2 - T_3) = m_{v2} (H_{v2} - H_{c2}) \tag{6.22}$$

Even after knowing the evaporator capacity and the desired concentration effect, it is difficult to compute the steam requirement from equation (6.15) through (6.19). In such cases, it is most logical to assume equal heat transfer areas for all evaporators in different steps which will result in the similar construction. The boiling point elevation and specific heat changes due to the concentration of the product may be overlooked while the effect of concentration of product on overall heat transfer coefficient should be computed for each effect. A plot of steam economy versus initial feed temperature (Fig.6.10) shows that for the normal temperatures, the steam economy varies between 2·0 and 2·5 and it gradually increases with the increase in feed solution temperature.

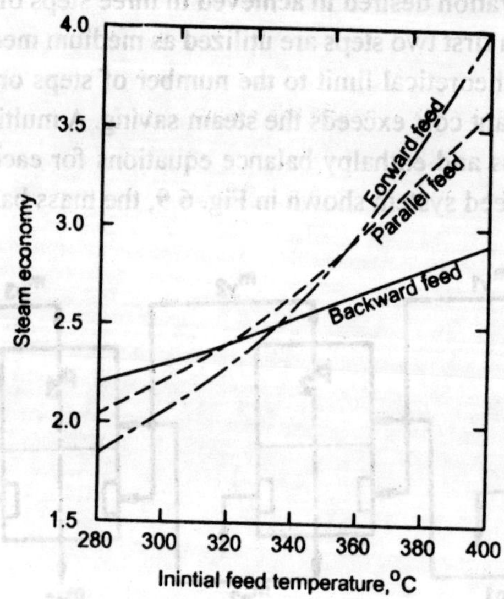

Fig. 6.10 *Economy of tripple-effect evaporator*

Methods of feeding

The most commonly used method of feeding the product through a multiple-effect evaporator is forward feed where the product fluid is pumped into the first effect and subsequently passed through the other effects [Fig. 6.11(*a*)]. It is the simplest form of flow and the fluid concentration increases from the first effect to the last. The first effect may be at atmospheric pressure or above, the second and further effects have, therefore, to be under increasingly lower pressures, or many a times under vacuum. This system requires pump for feeding a dilute liquid to the first effect and second pump to remove the thick liquor from the last effect, the transfer from effect to effect being done through control valves without pump due to the decreasing order of pressure.

Alternately, the feed may be passed in a reverse direction that is, starting at the last effect and then pumped through the successive effects to the first [Fig. 6.11(*b*)]. This is called as the backward feed method which requires pump between each pair of the effect because the flow is from low to high pressure with some pressure drop. It usually offers larger evaporation capacity than the forward feed system which may not be useful concerning to the product quality. When the product fluid is cold, it results in a lower economy than the forward feed method.

A mixed feed system [Fig. 6.11(*c*)] is also used where the dilute product fluid is fed at an intermediate effect which flows to the end effect by forward feed and is then pumped back to the first effect for final concentration. This system eliminates some of the pumps required in backward feed portion of the system and still allows the final evaporation at the highest temperature.

The dilute product fluid may also be admitted directly to each effect and concentrated thick liquor is taken out individually and there is no transfer of fluid product from one effect to the another. This is called as the parallel feed method [Fig. 6.11(*d*)], usually used in crystallizing evaporators where, slurry of crystals and mother liquor is withdrawn from each effect.

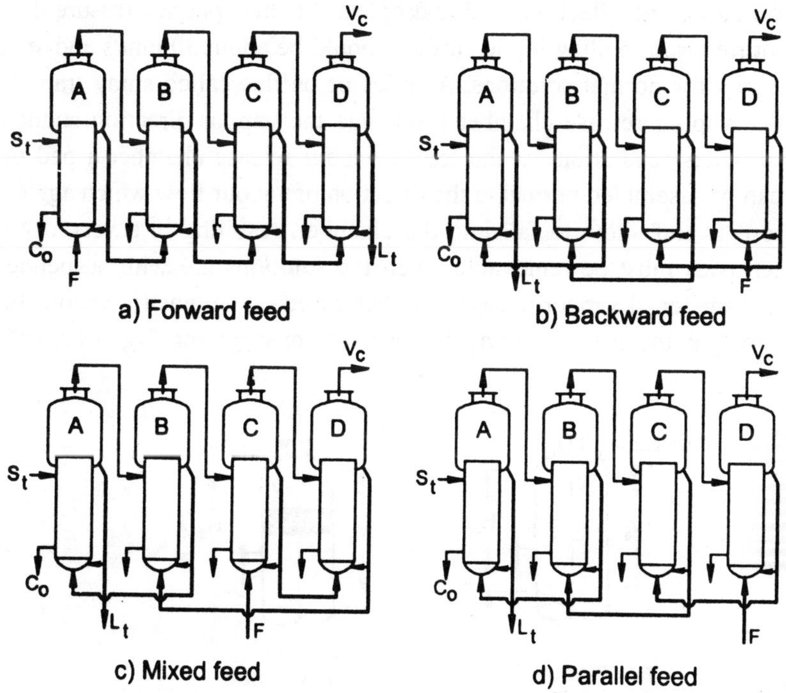

Fig. 6.11 *Feeding methods in multiple effect evaporator*

(C_o : condensate, F : feed, L_t : thick liquor, S_t : Steam, V_c : vapour to condenser)

In evaporators having large volume and where feed is mixed, the retention time of given food particle may be considerable. The retention time is obtained by dividing the evaporator volume by the feed rate but some part of fluid remains for longer period than this time. If the product is heat sensitive, its quality may be lowered. This problem can be avoided in modern high flow rate evaporators by the provision of a low hold up volume and there are very little or no chances for mixing of the fluid product. This consideration is required in plate evaporators, long tube evaporators and scraped plate.

6.7 Entrainment separators

During the vapour formation in evaporators and its movement towards the outlet, the liquid droplets of different sizes and quantity are entrained in the vapour. However, for heating purpose dry vapours are needed. Also the droplets moving out with vapours is a loss of product fluid. Hence, to separate the droplets from vapour the separators are provided. Following two methods are used for such an entrainment separation.

(*a*) *Vapour release chamber*

By locating a vapour release chamber just above the tube bundle shell or as a separate unit, the droplets settle down by gravity due to fall in velocity. Large droplets are separated by this method but has no effect on smaller droplets. At atmospheric pressure the vapour disengagement rate from boiling liquid surface should be about 30 mm/s and should not exceed 300 mm/s for normal solutions. A series of baffles called spray traps are also mounted in the vapour release chamber to change the vapour direction suddenly and enhance the droplets separation. At the top of vapour release chamber a pad of finely woven wire can be assembled normal to the direction of vapour flow which agglomerates the liquid droplets and allow to fall into the chamber, resulting in a very dry vapour. However, such pads must be removable when the solutions are with suspended solid particles. The diameter of vapour release chamber based on volumetric vapour flow rate and liquid foaming in the evaporator without entrainment separator [Fig. 6.12(*a*)] is given by,

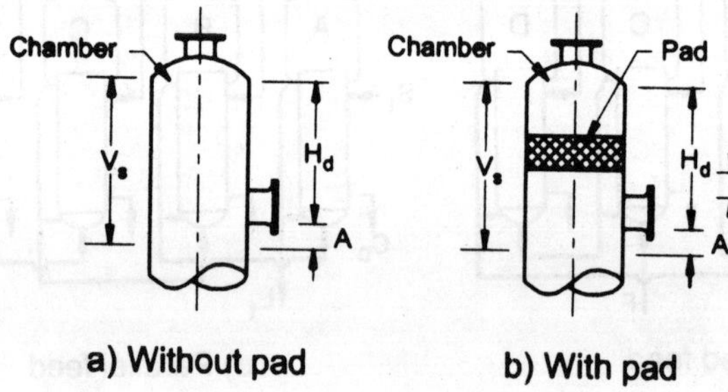

Fig. 6.12 *Drum with and without pad*

$$D_c = \left[\frac{148V}{\{(\rho_L - \rho_v)/\rho v\}^{0.5}}\right] \tag{6.23}$$

For drums with woven wire mash separator Fig. 6.12 (*b*)], the diameter is computed as,

$$D_c = \left[\frac{57V}{\{(\rho_L - \rho_v)/\rho_v\}^{0.5}}\right]^{0.5} \tag{6.24}$$

The disengaging height of vapour release chamber shown in [Fig. 6.12(*a*) and 6.12(*b*)] can be determined from the relationship shown by a graph (Fig. 6.13), with or without the entrainment separator. Vapour space extends minimum distance A (150 mm) below the inlet port and also above port in case of drum with pad. A shorter height can be provided where the separator forms an independent unit.

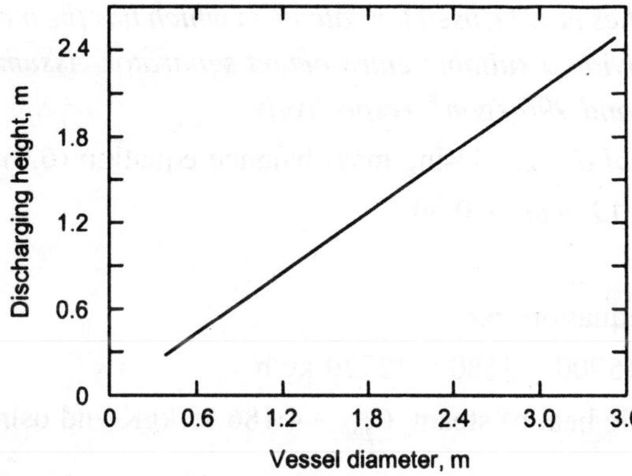

Fig. 6.13 *Relationship between drum diameter and disengaging height*

(b) Centrifugal separator

It consists of a separate cylinder with the helical flights provided around the outlet pipe (Fig. 6.14). Inlet pipe is provided so that the vapour enters tangentially and flows in a helical path through screw flights and the liquid droplets are separated in a process.

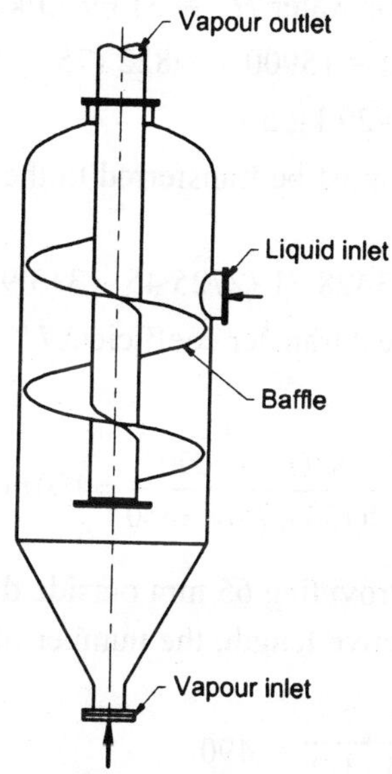

Fig. 6.14 *Centrifugal separator*

Example 6.1: *Orange juice is to be concentrated from 12 to 60% total solids @ 15·90 t/h at 28°C. The steam is available at 200 kPa gauge and evaporation chamber pressure is such that the juice evaporates at 75°C, the condensate being released at this temperature. Specific heat of juice and overall heat transfer coefficient may be assumed as 3·822 kJ/kg°C and 1271 W/m^2°C, respectively. The shell may be fabricated from low carbon steel sheet for which the modulus of elasticity is 1·84 (10)^5 N/mm^2 and allowable* *stress is 96 MPa. Brass tubes may be used for calendria which has the modulus of elasticity of 9·12 (10)⁴ N/mm². Provide a suitable entrainment separator. Assume the vapour and liquid densities as 0·125 and 996 kg/m³, respectively.*

Solution. (*a*) *Thermal design:* Using mass balance equation (6.9),

$$15{\cdot}90 \times 1000 \times 0{\cdot}12 = m_p \times 0{\cdot}60$$

or $m_p = 3180$ kg/h

Also with mass balance equation (6.8),

$$m_v = m_f - m_p = 15900 - 3180 = 12720 \text{ kg/h}$$

Assuming the specific heat of steam, $C_{pw} = 4{\cdot}186$ kJ/kgK and using equation (6.14) yields,

$$3{\cdot}822 = C_{ps} \times 0{\cdot}12 + 4{\cdot}186 \times 0{\cdot}88$$

or $C_{ps} = 1{\cdot}152$ kJ/kg °C

then $C_{pp} = 1{\cdot}152 \times 0{\cdot}6 + 4{\cdot}186 \times 0{\cdot}4 = 2{\cdot}365$ kJ/kg °C

From steam tables, the condensing steam temperature is 134°C at 200 kPa gauge pressure and enthalpy of steam, $H_{vs} = 2725{\cdot}45$ kJ/kg, at 75° C condensing temperature $H_v = 2321{\cdot}37$ kJ/kg and that for water $H_{cs} = 313{\cdot}9$ kJ/kg. Using equation (6.13),

$$m_s \times (2725{\cdot}45 - 313{\cdot}9) = 15900 \times 3{\cdot}822\,(75 - 28) + 12720 \times 2321{\cdot}37$$

or $m_s = 13428{\cdot}71$, say 13429 kg/h

The quantity of heat that must be transferred to the product for desired evaporation is,

$$q = m_s\,(H_{vs} - H_{cs}) = 13428{\cdot}71\,(2725{\cdot}45 - 313{\cdot}9) = 32384007 \text{ kJ/h}$$

Considering the overall heat transfer coefficient $U_o = 1271$ Wm²K, the area required for complete heat transfer is,

$$A = \frac{q}{U_0(T_s - T_e)} = \frac{32384007 \times 1000}{1271(134 - 75) \times 3600} = 120\,\text{m}^2$$

(*b*) *Mechanical design*: Providing 65 mm outside diameter brass tubes with 1·2 mm wall thickness and 1·2 m effective length, the number of tubes required is,

$$n = \frac{A}{\pi d_o L} = \frac{120}{\pi \times 0{\cdot}065 \times 1{\cdot}2} = 490$$

With 85 mm square pitch, the area occupied by the tubes is

$$a_n = S_T^2\, n = (0{\cdot}085)^2 \times 490 = 3{\cdot}54 \text{ m}^2$$

Area of downtake required @ 40% of the tube area is,

$$a_d = \frac{0{\cdot}4 \times 490\,\pi\,(0{\cdot}065)^2}{4} = 0{\cdot}65 \text{ m}^2$$

then $d = \left[\dfrac{4 \times 0 \cdot 65}{\pi}\right]^{0 \cdot 5} = 0 \cdot 90$ m

The total tube sheet area = 3·54 + 0·65 = 4·19 m^2

and $D_s = \left[\dfrac{4 \times 4 \cdot 19}{\pi}\right]^{0 \cdot 5} = 2 \cdot 30$ m

Alternately, assuming K_b = 1·17 for single tube pass, and using equation (5.19), the shell diameter for square pitch is,

$$D_s = 1 \cdot 128 \times 0 \cdot 085\ (490 \times 1 \cdot 17)^{0 \cdot 5} = 2 \cdot 30 \text{ m}$$

Orange juice is boiling at 75 °C, the corresponding pressure from steam table being 38·56 kPa (vacuum). Hence, considering the design pressure p = 80 kPa and allowable stress, f = 96 MPa, the calendria sheet thickness is given by equation (3.4).

$$t_s = \frac{80 \times 2300}{(2 \times 96 \times 1000 - 80)} = 0 \cdot 96 \text{ mm}$$

However, provide 6 mm thick calendria sheet for shell rigidity and corrosion allowance.

The tube sheet thickness can be designed similar to that of the shell and (fixed) tube exchanger using equation (5.20). To determine the coefficients in this equation, use equation (5.23) with D_o = 2312 mm.

$$K = \frac{1 \cdot 84(10)^5 \times 6(2312 - 6)}{9 \cdot 12(10)^4 \times 1 \cdot 2 \times 490(65 - 1 \cdot 2)} = 0 \cdot 744$$

Using equation (5.22) K_s for shell side pressure is,

$$K_s = \left[\frac{0 \cdot 744}{(2 + 3 \times 0 \cdot 744)}\right]^{0 \cdot 5} = 0 \cdot 419$$

Then, the tube sheet thickness is given by,

$$t_{ts} = 0 \cdot 419\ (2312 - 6)\left[\frac{0 \cdot 25 \times 80}{(96 \times 1000)}\right]^{0 \cdot 5} = 13 \cdot 95 \text{ mm}$$

With the corrosion allowance, provide 18 mm thickness for the tube sheet.

A conical cylindrical bottom is to be provided for the calendria with the following flange. t_f = 60 mm, number of bolts 100 (M12 × 1·5) D_p = 2360 mm and outside diameter as 2425 mm.

Providing a wire woven entrainment separator at the top of the tubes, the required evaporation chamber diameter can be determined by using equation (6.24) in which vapour volumetric flow is given by,

$$V = \frac{m_v}{\rho_v} = \frac{12720}{(0 \cdot 125 \times 3600)} = 28 \cdot 27 \text{ m}^3/\text{s}$$

Then $$D_c = \frac{7 \cdot 55(28 \cdot 27)^{0 \cdot 5}}{[(996 - 0 \cdot 125)/0 \cdot 125]^{0 \cdot 25}} = 4 \cdot 25 \text{ m}$$

Provide the shell height above tubes equal to 3·0 m.

The critical buckling pressure under vacuum or external pressure on cylindrical vessel is given by equation (3.26) with $t = 6$ mm.

$$p_c = K_c E \left(\frac{t}{D_o}\right)^c$$

In this case,

$$\frac{D_o}{L} = \frac{(2300 + 2 \times 6)}{1200} = 1 \cdot 93$$

From Tables 3.3, $K_c = 2 \cdot 253$ by interpolation and $c = 2 \cdot 54$. Then,

$$p_c = 2 \cdot 253 \times 1 \cdot 84\ (10)^5 \left(\frac{6}{2312}\right)^{2 \cdot 54} = 0 \cdot 112 \text{ N/mm}^2$$

Allowable external pressure can be determined using equation (3.28) where allowable stress in wall material is 96 MPa = 96 N/mm^2. Then,

$$p_e = 2 \times 96 \left(\frac{6}{2312}\right) = 0 \cdot 50 \text{ N/mm}^2$$

Thus, p_e is greater (about 5 times) than p_c, hence, the provided wall thickness is alright.

Providing a conical head both at top and bottom with $L = D_o$ and 6 mm thickness of wall, $K_c = 0 \cdot 879$ and $c = 2 \cdot 47$ for $D_o/L = 1 \cdot 0$.

$$p_c = 0 \cdot 879 \times 1 \cdot 84\ (10)^5 \left(\frac{6}{2312}\right)^{2 \cdot 47} = 0 \cdot 066 \text{ N/mm}^2$$

which is very small, hence, provided heads are alright.

Example 6.2: *A triple effect forward feed evaporator is to be provided for the conditions mentioned in Example 6.1, but with 16000 kg/h capacity, maintaining the boiling point temperature at 75°C, in the third effect of the system. Determine the steam requirement. Also compare the steam economy and the total heat transfer area required with a single effect system, assuming the respective heat transfer coefficients for the first, second and third stages as 2000, 1450 and 1050 W/m^2.*

Solution.

Given, t_s = 134 °C, t_f= 28 °C, t_3 = 75°C, U_1 = 2000 W/m^2 °C, U_2 = 1450 W/m^2°C, U_3 = 1050 W/m^2°C, Total temperature gradient = 134 – 75 = 59°C, m_f = 16000 kg/h = 4·44 kg/s, m_p = 16000 × 0·12/ 0·60 = 3200 kg/h = 0·89 kg/s , C_{pf} = 3·822 kJ/kg C, m_{v1} + m_{v2} + m_{v3} = m_f – m_p, C_{p2} = 2·80 kJ/kg °C, C_{p3} = 2·366 kJ/kg °C

Let, the temperature gradients of 19·5, 19·5 and 20 °C at stage 1, 2, and 3, respectively. Then the heat capacity (enthalpy) of steam and food product shall be as below.

Heat capacity of steam

Symbol	Gradient	Temperature	H_v, *kJ/kg*	H_c, *kJ/kg*
t_{sa}	--	134	2726·0	563·4
t_1	19·5	114·5	2698·3	480·4
t_2	19·5	95	2668·1	398·0
t_3	20	75	2635·3	313·9

Heat capacity of food product

Symbol	C_p	*Temperature*	*Heat content, kJ/kg*
H_f	3·822	134	475·1
H_{p1}	3·320	114·5	380·1
H_{p2}	2·800	95	266·0
Hp_3	2·366	75	177·5

Assuming equal heat transfer area at each stage, from equation (6.20), (6.21) and (6.22) yields,

$$\frac{q_1}{U_1(t_s - t_1)} = \frac{q_2}{U_2(t_1 - t_2)} = \frac{q_3}{U_3(t_2 - t_3)} \tag{1}$$

or

$$\frac{m_s(H_s - H_{c1})}{U_1(t_s - t_1)} = \frac{m_{v1}(H_{v1} - H_{c2})}{U_2(t_1 - t_2)} = \frac{m_{v2}(H_{v2} - H_{c3})}{U_3(t_2 - t_3)} \tag{2}$$

Putting the above values in equation (2),

$$\frac{m_s(2726 \cdot 0 - 480 \cdot 4)}{2000(134 - 114 \cdot 5)} = \frac{m_{v1}(2698 \cdot 3 - 398 \cdot 0)}{1450(114 \cdot 5 - 95)} = \frac{m_{v2}(2668 \cdot 1 - 313 \cdot 9)}{1050(95 - 75)}$$

Therefore, m_{v1} = 0·708 m_s, m_{v2} = 0·514 m_s, m_{v3} = 3·55 – 1·222 m_s

m_{p1} = 4·44 – 0·708 m_s

Using equation (6.12),

$$m_s\,(2726{\cdot}0 - 480{\cdot}4)$$

$$= [4{\cdot}44 - (4{\cdot}44 - 0{\cdot}708\, m_s)] \times 2698{\cdot}3 + (4{\cdot}44 - 0{\cdot}708\, m_s) \times 380{\cdot}1 - 4{\cdot}44 \times 475{\cdot}1$$

Then, $m_s = 2{\cdot}01$ kg/s

and $m_{v1} = 1{\cdot}423$ kg/s, $m_{v2} = 1{\cdot}033$ kg/s, $m_{v3} = 3{\cdot}55 - 1{\cdot}094$ kg/s

$$m_{p1} = m_f - m_{v1} = 4{\cdot}44 - 1{\cdot}423 = 3{\cdot}017 \text{ kg/s}$$

$$m_{p2} = m_f - m_{v1} - m_{v2} = 4{\cdot}44 - 1{\cdot}423 - 1{\cdot}033 = 1{\cdot}984 \text{ kg/s}$$

$$m_{p3} = m_f - m_{v1} - m_{v2} - m_{v3} = 4{\cdot}44 - 1{\cdot}423 - 1{\cdot}033 - 1{\cdot}094 = 0{\cdot}89 \text{ kg/s}$$

Considering heat balance,

$$m_{p2}\,H_{p2} + m_{v2}\,H_{v2} = m_{v3}\,H_{v3} + m_{p3}\,H_{p3} + m_{v2}\,H_{c3}$$

i.e., $1{\cdot}984 \times 266 + 1{\cdot}033 \times 2668{\cdot}1 = 1{\cdot}094 \times 2635{\cdot}3 + 0{\cdot}89 \times 177{\cdot}5 + 1{\cdot}033 \times 313{\cdot}9$

LHS = 3284 and RHS = 3365 *i.e.*, nearly equal, hence acceptable agreement. Hence, the steam requirement for the evaporator shall be,

(*a*) Tripple effect, $m_s = 2{\cdot}01 \times 3600 = 7236$ kg/h

Quantity of water evaporated = 16000 – 3200 = 12800 kg

i.e., $SE_3 = 12800/\ 7236 = 1{\cdot}769$ kg water/ kg steam

(*b*) Single effect, $SE_1 = 12720\ /13429 = 0{\cdot}947$ kg water/kg steam

Hence, the enhanced steam economy under the similar conditions but the tripple effect is,

$$SE = 100 \times (1{\cdot}769 - 0{\cdot}947)/\ 0{\cdot}947 = 86{\cdot}80\%$$

Heat transfer area can be computed as below using a heat balance equation at any stage of the three.

$$A_1 = \frac{m_s(H_s - H_c)}{U_1(t_s - t_1)} = \frac{2{\cdot}01(2726{\cdot}0 - 480{\cdot}4) \times 1000}{2000(134 - 114{\cdot}5)} = 115{\cdot}73 \text{ m}^2$$

The total area required is,

$$A_3 = 3 \times 115{\cdot}73 = 347{\cdot}19 \text{ m}^2$$

Area required for single effect is,

$$A_1 = 120 \times 12800/\ 12720 = 120{\cdot}75 \text{ m}^2$$

Therefore, an increase in heat transfer rate due to tripple effect is,

$$A_{\%} = 100 \times (347{\cdot}19 - 120{\cdot}75)/\ 120{\cdot}75 = 187{\cdot}53\ \%.$$

Nomenclature

A_m : average area on product and steam sides, m^2
A_p : product (shell) side area of heat transfer, m^2
A_s : steam (tube) side area of heat transfer, m^2
A_1 : heat transfer area at stage 1, m^2
A_2 : heat transfer area at stage 2, m^2
A_3 : heat transfer area at stage 3, m^2
a_d : required area of down take, m^2
C_{pf} : specific heat of feed solution, kJ/kg °C
C_{pp} : specific heat of product, kJ/kg °C
C_{ps} : specific heat of steam, kJ/kg °C
C_{pw} : specific heat of water, kJ/kg °C
c_1, c_2 : linearity constant
D_c : diameter of vapour release chamber, m
d : diameter of the down take, m
D_s : evaporator shell diameter, m
g : gravitational constant, m/s^2
H_{cs} : heat content (enthalpy) of condensed steam, kJ/kg
H_{c1} : heat content (enthalpy) of vapour / steam at stage 1, kJ/kg
H_{c2} : heat content (enthalpy) of vapour / steam at stage 2, kJ/kg
H_f : heat content (enthalpy) of feed solution, kJ/kg
H_p : heat content (enthalpy) of product, kJ/kg
H_{p1} : heat content (enthalpy) of product at stage 1, kJ/kg
H_{p2} : heat content (enthalpy) of product at stage 2, kJ/kg
H_{p3} : heat content (enthalpy) of product at stage 3, kJ/kg
H_v : heat content (enthalpy) of vapour, kJ/kg
H_{vs} : heat content (enthalpy) of steam in vapour state, kJ/kg
H_{v1} : heat content (enthalpy) of vapour at stage 1, kJ/kg
H_{v2} : heat content (enthalpy) of vapour at stage 2, kJ/kg
H_{v3} : heat content (enthalpy) of vapour at stage 3, kJ/kg
h_p : surface heat transfer coefficient on product side, kJ/m °C
h_s : surface heat transfer coefficient on steam side, kJ/m °C
K : coefficient depending on material and geometry of shell and tube
K_f : constant
K_s : multiplying factor for tube sheet thickness
k : thermal conductivity of material, kJ/h m °C

L : length of heating surface, mm
M : rate of condensate collection, kg/h
m : number of fractions from screens
m_f : mass flow rate of feed solution, kg/h
m_p : mass flow rate of product, kg/h
m_{p1} : mass flow rate of product at stage 1, kg/h
m_{p2} : mass flow rate of product at stage 2, kg/h
m_{p3} : mass flow rate of product at stage 3, kg/h
m_v : mass flow rate of vapour, kg/h
m_{v1} : mass flow rate of vapour at stage 1, kg/h
m_{v2} : mass flow rate of vapour at stage 2, kg/h
m_{v3} : mass flow rate of vapour at stage 3, kg/h
m_s : mass flow rate of steam, kg/h
p : vapour pressure of pure water, N/mm^2
P' : vapour pressure of liquid product containing solids, N/mm^2
q_1 : rate of heat transfer at stage 1, kJ/h
q_2 : rate of heat transfer at stage 2, kJ/h
q_3 : rate of heat transfer at stage 3, kJ/h
R : universal gas constant
SE : steam economy, %
SE_1 : steam economy for single effect, kg water/kg steam
SE_3 : steam economy for tripple effect, kg water/kg steam
T_a : boiling point of pure water, °C
T_a' : boiling point of liquid product containing solids, °C
T_e : evaporator temperature, °C
T_f : temperature of feed solution, °C
T_s : temperature of condensing steam, °C
T_w : temperature of shell wall, °C
T_1 : temperature of product at stage 1, °C
T_2 : temperature of product at stage 2, °C
T_3 : temperature of product at stage 3, °C
U_o : overall heat transfer coefficient, kJ/m^2 °C
U_1 : overall heat transfer coefficient at stage 1, kJ/m^2 °C
U_2 : overall heat transfer coefficient at stage 2, kJ/m^2 °C
U_3 : overall heat transfer coefficient at stage 3, kJ/m^2 °C
V : volumetric flow rate of vapour, m^3/h

X_f	: mass or mole fraction of feed solution
X_p	: mass or mole fraction of product
X_s	: mass or mole fraction of steam
X_w	: mass or mole fraction of water
μ_f	: viscosity of the fluid, Ns/m^2
ρ_f	: density of fluid, kg/m^3
ρ_L	: density of liquid, kg/m^3
ρ_v	: density of vapour, kg/m^3
λ_s	: latent heat of steam, kJ/kg
λ_v	: latent heat of vapour, kJ/kg
λ_w	: latent heat of water vaporization, kJ/kg
λ_p	: latent heat of liquid product containing solids, kJ/kg
θ_1	: temperature gradient at stage 1, °C
θ_2	: temperature gradient at stage 2, °C
θ_3	: temperature gradient at stage 3, °C

7

Design of Vessel Supports

7.1 Introduction

To complete the design of any process vessel consideration and selection of its supports is essential. Based on the process requirement, the vessel may be supported either in vertical or horizontal position. It is necessary to analyse the stresses induced due to supports on vessel walls or other components so as to make the installation strong, safe and cost effective. Vertical vessels are commonly supported by lug or bracket, skirt, leg and ring type of supports whereas the horizontal vessels are provided with the saddle, leg and ring type supports. The selection criteria for the supports are dependent on the size, shape and weight of the vessel, operating pressure and temperature, availability of floor area, room head, convenience of location, construction material, and location of various connections.

7.2 Lug or bracket supports

These are most commonly used for vertical vessels with height to diameter ratio 2 to 3. They can be easily fabricated out of plates and fixed on the vessels with minimum length of welding. The brackets rest on the columns or structural beams. These supports are most suitable for thick walled vessels so as to absorb bending, compressive, tensile and shear stresses induced in the vessel walls. Thin walls require additional reinforcement if brackets are provided. They can take care of the dimensional thermal expansions, if sliding arrangement is provided. Lug or bracket supports are easily levelled and attached to the vessel wall with minimum weld length and are comparatively cheaper with other types of supports. Two brackets are necessary upto 0·8 m diameter, four upto 3·0 m, 6 upto 5·0m and 8 for vessels with diameter above 5·0 m. Figure 7.1 shows the details of bracket or lug and the forces on supported vessel wall. The maximum compressive load on bracket due to wind load, vessel self weight and the weight of contents is the controlling load on lee-ward side of the vessel for bracket design, given by.

$$P = \frac{4P_w(H_1 - H_2)}{n_b D_p} + \frac{(W_1 + W_2)}{n_b} \tag{7.1}$$

When the pressure vessel is installed indoor, $P_w = 0$ and design pressure should be equal to the second term in equation (7.1) only.

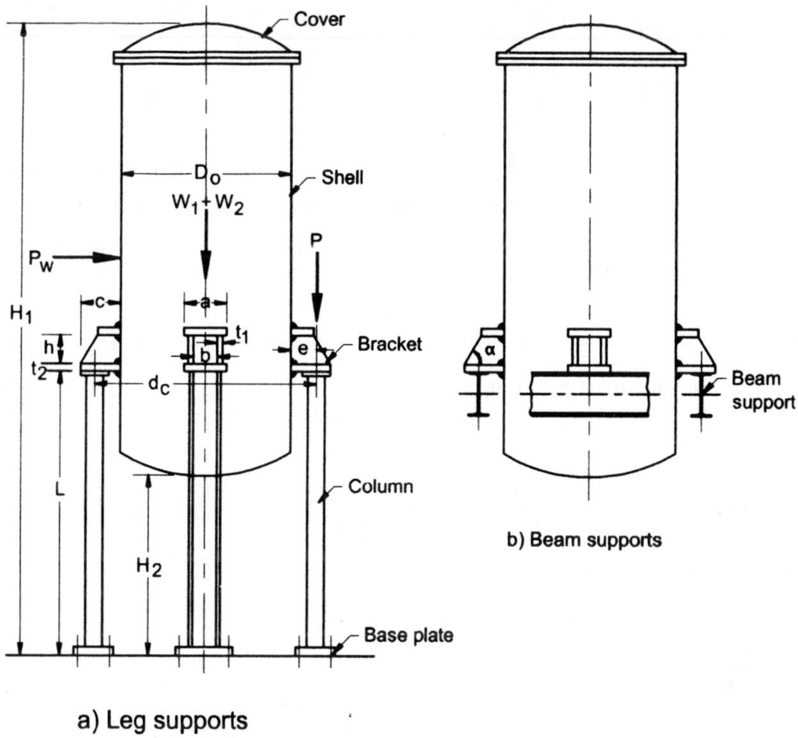

Fig. 7.1 *Bracket support for vertical vessel*

The bracket plates carry vertical shear load, consisting of three plates, two vertical members called gusset plates and one horizontal base plate called the stiffener. The stiffener may be two in numbers, additional is provided at the top of the gusset plates. In some cases, both the stiffeners may be absent. Three plate brackets are commonly used. The bottom stiffener or base plate is fixed with vessel shell on the edges with the load distributed over about half of its area. Average pressure on base plate is given by,

$$P_a = \frac{P}{bc} \tag{7.2}$$

and the maximum stress is equal to,

$$f_a = 0{\cdot}5\ p_a \frac{c^2}{t_{st}^2}\left[\frac{b^2}{c^2+b^2}\right] \tag{7.3}$$

This stress is distributed on surface of contact between the stiffener and the supporting beam, the actual stress may be considered 40% higher than given by equation (7.3), then,

$$f_a = 0{\cdot}7\ p_a \frac{c^2}{t_{st}^2}\left[\frac{b^2}{c^2+b^2}\right] \tag{7.4}$$

The stiffener thickness shall be chosen such that the stress f_a should be within the permissible stress limit. Two gusset plates are provided at each support. Hence, the bending moment for each plate with the eccentricity equal to the dimension *a* (Fig. 7.1) is,

$$M = 0{\cdot}5\ P\ e$$

The section modulus of plate is,

$$Z_p = \frac{t_g h^2}{6}$$

and the stress at the edge is,

$$f_e = \frac{3Pe}{t_g h^2} \tag{7.5}$$

The edge is at an angle $\propto$ from the horizontal. Therefore, the maximum compressive stress parallel to the edge of the gusset plates is,

$$f_e = \frac{3Pe}{(t_g h^2 \cos \propto)} \tag{7.6}$$

The gusset plate thickness should be so chosen that the stress shall not exceed the permissible limit. Table 7.1 gives major dimensions of lug or bracket supports used in practice.

Table 7.1 Approximate dimensions for brackets

No. of supports	Width of stiffener (c), mm	Height of gusset plate (a), mm	Plate thickness	
			Stiffener	Gusset
2	150	140	10	8
3	150	140	10	8
4	150	140	10	8
5	200	180	13	10
6	200	180	13	10
7	200	180	13	10
8	200	180	13	10

Columns support the brackets and the load on column is not concentric with the total weight of the vessel and its contents. These loading conditions produce an axial compression and also bending due to eccentric loading. If total load (P) is acting on a short column ($l/r < 250$), the maximum combined direct and bending stress is given by,

$$f = \frac{P}{(A_c n_c)} + \frac{Pe}{(n_c Z)} \tag{7.7}$$

The columns should be provided such that the effective column length, $l_e/r < 180$.

The bending stress due to wind load is given by,

$$f_w = \left(\frac{P_w}{n}\right)\left(\frac{L_c}{2Z}\right) \tag{7.8}$$

When the columns are subjected to the direct and bending loads simultaneously, the following condition must be satisfied.

$$\frac{P / A_c}{f_c} + \frac{f_{ce} + f_w}{f_t} \leq 1 \cdot 0 \tag{7.9}$$

7.3 Skirt support

Skirt of shell support is a conical [Fig. 7.2(a)] or cylindrical [Fig. 7.2(b)] shell welded to the base of the vessel. A flange at the bottom of the skirt transmits the load to the foundation. Normally, reinforced openings are provided in the skirt for access and for the connecting pipes. The skirt may be welded either to the bottom head of the process vessel as shown in Fig. 7.3 or to the outside surface of the shell [Fig.7.3(a)], or flush with shell [Fig. 7.3(b)] or to the bottom head of the vessel [Fig. 7.3(c)], the flush with shell arrangement being most commonly used. Skirt support does not impose concentrated load on vessel shell and hence, are recommended for vertical vessels and tall columns subjected to wind load. A flat ring with or without gussets is provided as bearing plate in contact with the concrete foundation.

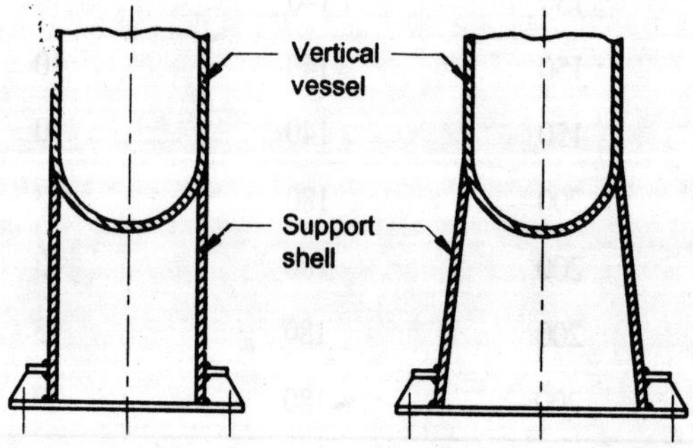

(*a*) Cylindrical shell (*b*) Conical shell

Fig. 7.2 *Types of skirt support*

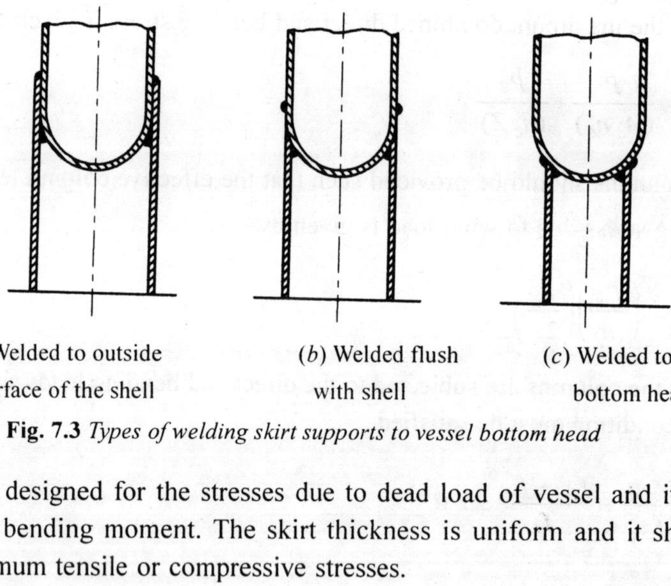

(*a*) Welded to outside surface of the shell (*b*) Welded flush with shell (*c*) Welded to the bottom head

Fig. 7.3 *Types of welding skirt supports to vessel bottom head*

The skirt is designed for the stresses due to dead load of vessel and its contents, wind or seismic bending moment. The skirt thickness is uniform and it should stand against the maximum tensile or compressive stresses.

(*a*) Stress due to dead loads (f_d) is,

$$f_d = \frac{W}{(\pi D_{sk} t_{sk})}$$

(*b*) Stress due to wind load: Wind forces are acting horizontally on vessel resulting in overturning moment in the skirt. Wind force acting on lower region upto 20 m height is given by,

$$P_{lw} = K\, p_1\, h_1\, D_o \qquad [7.11(a)]$$

and above 20 m height is,

$$P_{uw} = K\, p_2\, h_2\, D_o \qquad [7.11(b)]$$

The respective ranges for wind pressure for lower and upper parts of the vessel are 40 to 100 kg/m^2 and 100 to 200 kg/m^2. The coefficient of shape factor is taken as 0·70 for cylindrical surface.

Bending moments at the base of vessel due to wind can be determined by,

$$M_{lw} = P_{lw} \frac{h_s}{2} \text{ (up to 20 m height)} \qquad [7.12(a)]$$

and $M_{uw} = P_{lw} (0{\cdot}5\, h_1) + P_{uw} [h_1 + (0{\cdot}5\, h_2)$ (above 20m height) $\qquad [7.12(b)]$

The stress is given by,

$$f_{wb} = \frac{M_w}{Z} = \frac{4\, M_w}{\pi\, D_{sk}^2\, t_{sk}} \qquad (7.13)$$

The maximum tensile stress in the skirt wall is,

$$f_{te} = f_{wb} - f_d \qquad (7.14)$$

and the maximum compressive stress is,

$$f_{cs} = f_{wb} + f_d \qquad (7.15)$$

It is usual practice to assume the suitable wall thickness (t_{sk}) and check for the maximum stresses, which should not exceed the permissible values of the materials.

The maximum compressive stress between the bearing plate and concrete surface is given by,

$$f_c = \frac{W}{A_f} + \frac{M_w}{Z} \qquad (7.16)$$

In this case, f_c must be less than the allowable stress in the concrete. The thickness of bearing plate can be determined by considering it as a uniform cantilever beam with f_c as minimum load,

$$M_{max} = 0{\cdot}5\, f_c\, b_r\, l_r^2$$

The maximum stress in an elemental strip of unit width (b = 1) will be,

$$f_{max} = \frac{6 M_{max}}{b t_B^2} = \frac{3 f_c b_r l_r^2}{b t_B^2} = \frac{3 f_c l_r^2}{t_B^2} \qquad (7.18)$$

solving equation (7.18) for t_B gives,

$$t_B = b_a \left[\frac{3f_c}{f_{max}}\right]^{0.5} \tag{7.19}$$

If equation (7.19) gives the value after rounding to 12 mm or less, a steel rolled angle may be used for the bearing plate, thickness of which is given by,

$$\frac{3f_c(b_a - 1.7t_a^2)}{t_a^2} \le f \tag{7.20}$$

If the determined plate thickness is between 12 to 20 mm, a single ring bearing plate may be provided. If gusset plates are provided, it will strengthen the bearing plate resulting in the reduction in its thickness.

The anchor bolts in concrete will provide the resistance to overturning of the vessel for wind or seismic forces and are subjected to the tensile loads only. The minimum stress between the bearing plate and concrete foundation will be,

$$f_{cb} = \frac{W_1}{A_f} - \frac{M_w}{Z} \tag{7.21}$$

where, W_{min} is the self weight of empty vessel only.

If $f_c \ge 0$, it is required to determine the so-called coefficient of stability as below,

$$K_s = \frac{M_d}{M_w} = \frac{W_1 R_m}{M_w}$$

The moment arm for W_{min} is equal to 0·42 times the outer diameter of bearing plate. If the value of K_s of works out to be greater than 1·5, there is no need for anchorage of the vessel or if bolts are provided, they have a function only to hold the vessel in position. If $K_{s(min)}$ gives a negative value the vessel must be anchored. The load on each bolt is given by,

$$P_{bolt} = \frac{f_{cb} A_f}{N} \tag{7.22}$$

The minimum number of bolts is taken as 1·67 D_{sk}, but actual number of bolts shall be multiple of four, preferably the minimum bolt diameter being not less than 24 mm. The stress in bolt is,

$$f_{bolt} = \frac{P_{bolt}}{A_b} \tag{7.23}$$

The above stress given by equation (7.23) should not exceed the permissible tensile stress in the bolt material.

7.4 Saddle supports

Generally, horizontal cylindrical vessels are supported on saddles, placed at two positions (Fig. 7.4). It can be more, if unavoidable with stiffening rings. The location of the saddle supports should be decided with the consideration of advantage of stiffening due to head. For large thin-walled vessels, or vessels under vacuum should be provided with ring supports. For small vessels, simple leg supports are used. The saddle supports are located at a distance $A = 0{\cdot}4\ R$ from the tangent line of vessel shell (Fig. 7.4) and not exceeding $0{\cdot}20\ L$, where R and L are the radius and length of vessel shell, respectively. The included angle, θ of a saddle support should not normally be less than 120°.

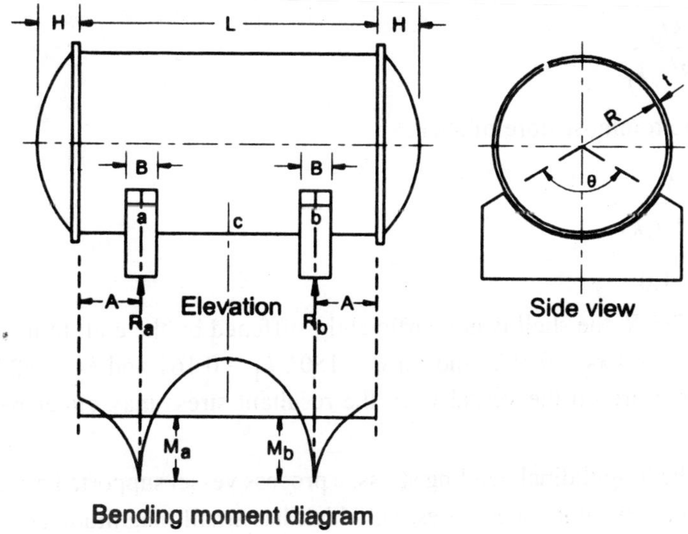

Fig.7.4. *Horizontal cylindrical vessels mounted on saddle supports.*

A cylindrical vessel with hemispherical or elliptical head may be treated as an equivalent cylinder having the length equal to $(L + 1{\cdot}33\ H)$ where, H is the depth of the dished head. Horizontal vessel is assumed to act through the centre of the vessel. Then the reaction at each support is,

$$R_a = R_b = 0{\cdot}5\ w\ (L + 1{\cdot}33\ H) \quad (7.24)$$

The maximum bending moments induced at the supports and at the centre of the vessel are calculated by using the following relationship.

At the supports,

$$M_a = R_a\,A\left\{1 - \frac{1-(A/L)+[(R^2-H^2)/2AL]}{1+(1{\cdot}33\,H/L)}\right\} \quad (7.25)$$

At the centre of the span,

$$M_c = \frac{R_a L}{4}\left\{\frac{1+[2(R^2-H^2)/L^2]}{1+(1\cdot 33H/L)} - \frac{4A}{L}\right\} \tag{7.26}$$

The longitudinal bending stress in the shell will depend on the stuffiness of the shell at the cross section where the bending is maximum.

(*a*) In case the stiffness is enough to maintain a circular cross section (*i.e.*, $A < 0{\cdot}5R$), the whole cross section of the vessel is effective and stress due to bending is given by,

(*i*) At the topmost fibre of shell,

$$f_{to} = \frac{M_a}{\pi R^2 t_s k_1} \tag{7.27}$$

(*ii*) At the bottommost fibre of shell,

$$f_{bo} = \frac{M_a}{\pi R^2 t_s k_2} \tag{7.28}$$

where, $k_1 = k_2 = 1.0$.

(*b*) For $A > 0{\cdot}5\,R$, the shell is not sufficiently stiffened by the end. In this case for $\theta = 120°$, $k_1 = 0{\cdot}107$ and $k_2 = 0{\cdot}192$, and for $\theta = 150°$, $k_1 = 0{\cdot}161$ and $k_2 = 0{\cdot}279$. If there is any internal pressure on the vessel wall, the resultant stress may either be tensile or compressive.

In addition the longitudinal bending stress, a process vessel supported on saddles will be subjected to the tangential shear stresses, which has small magnitude and may be of limited value if the shell thickness is greater than $0{\cdot}1\,R$ and $A < 0{\cdot}5\,R$.

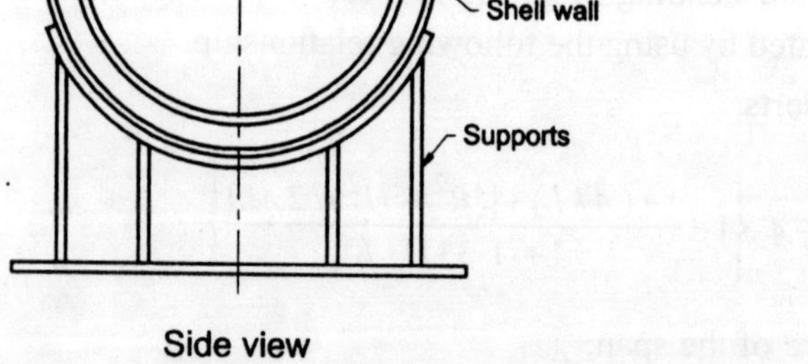

Fig. 7.5 *Stiffener ring inside the shell*

The stress in the shell wall at mid-span is given by

$$f_{ms} = \frac{M_c}{\pi R^2 t_s} \quad (7.29)$$

which may either be tensile or compressive depending on the position of the fibre. The resultant tensile stresses including axial stress due to internal vessel pressure should not exceed the permissible stress and the resultant compressive stress should not exceed the permissible compressive stress or (t_s E/ 16 R) whichever is less.

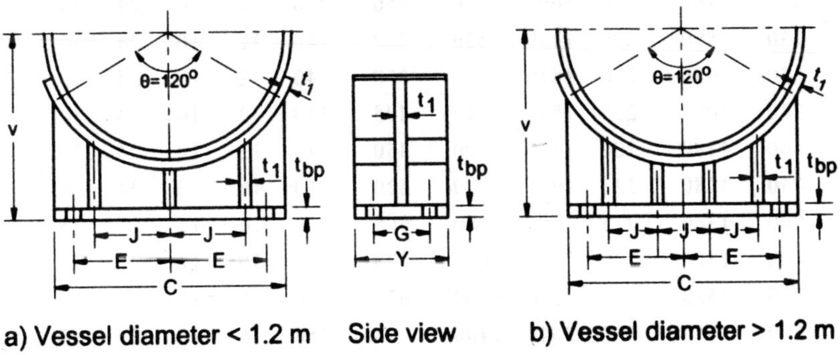

Fig. 7.6 *Steel saddles and standard dimensions*

The shell alone is not sufficient to resist the circumferential bending moment in case of $A > 0{\cdot}5\ R$ and thin-walled process vessels. Therefore, ring stiffeners are provided (Fig. 7.5) to the shell, usually inside, in order to share the shell wall load. Ring stiffener is designed from the following equation.

$$f_{ac} = \frac{K' R_a}{A} \pm \frac{K'' R_a R}{Z_r} \quad (7.30)$$

where, $K' = 0{\cdot}056$, $K'' = 0{\cdot}053$ for $\theta = 120°$ and $K' = 0{\cdot}021$ $K'' = 0{\cdot}032$ for $\theta = 150°$.

The saddles must be designed to take all the load from vessel, its contents and connections. They may be constructed of bricks or concrete or are fabricated from steel plates. The contact angle θ should be 120° and 150°. Wear plates are provided as reinforcement, welded to the shell wall over the area of contact with saddle. One saddled for each vessel should be provided with slot holes for sliding under movement due to thermal expansion such as heat exchangers or to axial strain in a long vessel are expected. Such arrangement is not required in short (< 3 m length) cold process vessels. Table 7.2 with reference to Fig. 7.6 gives the standard dimensions for saddles used practice.

Table 7.2 Dimension of welded saddles for horizontal process vessels

Vessel dia., mm	Max. in operation, kN	Saddle dimensions according to notations in Fig. 7.6 (mm)											
		V	*Y*	*C*	*E*	*J*	*G*	*tbp*	*tl*	*Bolt size*	*Bolt holes*	*Slotted holes*	*Fillet weld*
600	35	480	150	550	240	190	95	6	5	20	25	25×37	5
800	50	580	150	700	290	225	95	8	5	20	25	25×37	5
900	65	630	150	810	340	275	95	10	6	20	25	25×37	5
1000	90	680	150	910	390	310	95	11	8	20	25	25×37	5
1200	180	780	200	1090	450	350	140	12	10	24	30	30×45	6
1400	230	880	200	1240	530	350	140	12	10	24	30	30×45	6
1600	330	980	200	1410	620	350	140	12	10	24	30	30×45	6
1800	380	1080	200	1590	710	405	140	12	10	24	30	30×45	6
2000	460	1180	200	1770	800	450	140	12	10	24	30	30×45	6
2200	750	1280	225	1950	890	520	150	16	12	24	30	30×45	6
2400	900	1380	225	2130	980	565	150	16	12	27	33	33×52	10
2600	1000	1480	225	2300	1030	590	150	16	12	27	33	33×52	10
2800	1350	1580	250	2500	1100	625	150	16	12	27	33	33×52	10
3000	1750	1680	250	2640	1180	665	150	16	12	27	33	33×52	10
3200	2000	1780	250	2820	1260	730	150	16	12	27	33	33×52	10
3600	2500	1980	250	3200	1400	815	150	16	12	27	33	33×52	10

Note: *The maximum allowable working stress is assumed 95 MN/mm² and the welding should be continuous fillet weld all around the joints.*

Example 7.1: *A process vessel of 2000 mm diameter and 2·4 m height is to be provided with four leg supports at a height of 2·4 m above foundation. The clearance of vessel bottom to the foundation is 1·0 m and the bolt circle diameter is 2150 mm. The loads due to self weight of vessel and its contents are 10250 N and 25250 N, respectively. If the wind load is 1250 N/m² on vessel and allowable stresses in compression and bending are 125 and 160 N/mm², respectively design the leg supports.*

Solution. From equation [7.11 (*a*)],

$$P_w = 0{\cdot}7 \times 1250 \times 2{\cdot}4 \times 2{\cdot}0 = 4200 \text{ N}$$

Load on lee-ward side on bracket (leg) from equation (7.1) is,

$$P = \frac{4 \times 4200(2400 + 1000 - 1000)}{4 \times 2150} + \frac{10250 + 25250}{4} = 13563 \text{ N}$$

Referring to Table 7.1, b = 140 mm and c = 150 mm for four supports. The average pressure on base plate will be,

$$p_a = \frac{13563}{140 \times 150} = 0{\cdot}65 \text{ N/mm}^2$$

and the maximum stress from equation (7.4) will be,

$$f_a = 0{\cdot}7 \times 0{\cdot}65 \times \frac{(150)^2}{(t_{st})^2}\left[\frac{(140)^2}{(150)^2 + (140)^2}\right] = 160 \text{ N/mm}^2$$

Therefore, t_{st} = 5·46 mm. Hence, provide 6 mm thick plate for stiffener. Max compressive stress parallel to gusset plate from equation (7.6) a = 25 mm and h = 140 mm assuming $\propto$ = 45 °C,

$$f_c = \frac{3 \times 13563 \times 75}{t_g (140)^2 \times 0{\cdot}707} = 125$$

then, t_g = 1·76 mm, with corrosion allowance of 2 mm, provide 4 mm gusset plates.

It is proposed to provided ISMC 100 channel section for columns.

Then A_c = 1220 mm^2, r_y = 14·8 mm, Z_y = 7710 mm^3 from standard tables.

Effective length l_e = 0·65 × 2400 = 1560 mm for both ends held in position and restrained then, slenderness ratio, $\frac{l_e}{r} = \frac{1560}{14 \cdot 8}$ = 105·40 *i.e.*, < 180 OK.

Allowable stress for ratio 105·40, f_a = 74·68 N/mm^2.

The combined stress from equation (7.7) will be,

$$f = \left(\frac{13563}{4 \times 1220}\right) + \left(\frac{13563 \times 75}{4 \times 7710}\right)$$

$$= 2{\cdot}78 + 32{\cdot}98$$

$$= 35{\cdot}76 \text{ N/mm}^2$$ *i.e.*, within the limit

Check $\left(\frac{2 \cdot 78}{125}\right) + \left(\frac{32 \cdot 98}{160}\right)$ = 0·2685 *i.e.*, < 1·0 hence, the design is alright.

Tying ISMC 75 results unsafe and ISMC 125 will be uneconomical.

Example 7.2: *A 25 m tall vessel weighting 1·03 MN with 150 kN product in it, is to be installed having 2500 mm diameter and is subjected to 1300 N/m2 wind pressure. If a straight vertical skirt with the same vessel diameter and 4·5 m height is to be provided, design the skirt support for proposed vessel. Assume the tensile and compressive stresses in the skirt material as 125 and 65 N/mm^2, respectively.*

Solution. Total load of vessel and contents = 1·03 + 150 $(10)^{-3}$ = 1·18 MN

Stress due to dead load from equation (7.10) is,

$$f_d = \frac{1 \cdot 18 (10)^6}{2500 \pi t_{sk}} = \frac{150 \cdot 24}{t_{sk}}$$

Wind forces on lower and upper regions from equation (7.11) are,

$$p_{lw} = 0{\cdot}7 \times 1300 \times 20 \times 2{\cdot}5 \times (10)^{-3} = 45{\cdot}5 \text{ kN}$$

and $$p_{uw} = 0{\cdot}7 \times 1300\ (5 + 4{\cdot}5) \times 2{\cdot}5\ (10)^{-3} = 21{\cdot}61 \text{ kN}$$

Respective bending moments will be,

$$M_{lw} = \frac{45\cdot5\times 20}{2} = 455 \text{ kNm}$$

and $M_{uw} = 455 + 21\cdot61\,(20 + 0\cdot5 \times 9\cdot5) = 989\cdot85$ kNm

Then the stresses due to wind from equation (7.13) will be,

$$f_{wb} = \frac{4[455+989\cdot85](10)^6}{\pi(2500)^2 t_{sk}} = 294\cdot34/\,t_{sk}$$

The maximum tensile stress in skirt wall is,

$$f_{te} = \frac{294\cdot34}{t_{sk}} - \frac{150\cdot24}{t_{sk}} = \frac{144\cdot10}{t_{sk}} = 125$$

hence, $t_{sk} = \dfrac{144\cdot10}{125} = 1\cdot15$ mm

and the maximum compressive stress is,

$$f_{cs} = \frac{294\cdot34}{t_{sk}} - \frac{150\cdot24}{t_{sk}} = \frac{444\cdot58}{t_{sk}} = 65$$

hence, $t_{sk} = \dfrac{444\cdot58}{65} = 6\cdot84$ mm

Considering the corrosion, allowance, provide 100 mm thick plate for fabrication of skirt.

Let the bolt circle diameter of skirt bearing plate be D_{sk} + 300 mm = 2500 + 300 = 2800 mm. Then the maximum compressive stress on bearing plate from equation (7.16) is,

$$f_c = \frac{1\cdot18(10)^6\times 4}{[(2800)^2-(2500)^2]\pi} + \frac{0\cdot7\times1300\times25\times100(25+4\cdot5)^2/2}{[(2800)^4-(2500)^4]\pi/(32\times2800)}$$
$$= 0\cdot94 + 1\cdot26$$
$$= 2\cdot20 \text{ N/mm}^2 \; i.e., < 4\cdot0 \text{ N/mm}^2 \text{ for M 15 concrete mix}$$

Maximum bending moment in bearing plate from equation (7.17)

$$M_{\max} = \frac{2\cdot20\times1\cdot0(300/2)^2}{2} = 24750 \text{Nmm}$$

and $f_{max} = \dfrac{6\times24750}{1\times t_B^2} = \dfrac{1\cdot485(10)^5}{t_B^2} = 160$ N/mm^2

then $$t_B = \left[\frac{1 \cdot 485(10)^5}{160} \right]^{0.5} = 30{\cdot}50 \text{ mm}$$

As $t_B > 20$ mm, provide a bolting chair.

Minimum stress on bearing plate will be,

$$f_{cb} = \frac{1 \cdot 03(10)^6 \times 4}{[(2800)^2 - (2500)^2]\pi} - 1{\cdot}26 = -0{\cdot}435 \text{ N/mm}^2$$

Since, this stress is negative, the vessel skirt must be anchored through the bolts to the concrete foundation. Assuming 20 bolts, from equation (7.22).

$$P_{\text{bolt}} = \frac{0 \cdot 435}{20} \cdot \frac{\pi}{4} [(2800)^2 - (2500)^2] = 27161 \text{ N}$$

With allowable stress in bolt as 57·3 N/mm² from Table (3.8) for hot rolled carbon steel bolts,

$$A_b = \frac{22478}{57 \cdot 3} = 474{\cdot}01 \text{mm}^2$$

Hence, provide 20 – 30mm diameter bolts for which the effective cross sectional area is,

$$A_e = \left(\frac{\pi}{4}\right) 30 - 2 \times (2)^2 = 530{\cdot}93 \text{ mm}^2\text{, OK}$$

Example 7.3: *A horizontal vessel of shell length 7·2 m and 1200 mm diameter is mounted on two saddle supports spaced 3·32 m away from centre towards both ends. Vessel has torispherical heads having crown radius of 1220 mm and total head depth of 250 mm. Allowing 2 mm for corrosion, the shell and head thickness is 10 and 12 mm, respectively. The internal pressure in the vessel is 0·45 N/mm2 and it weight 103·2 k/N with its contents. Check the stresses in the saddle supports, if the allowable stress in the shell material is 92 N/mm².*

Solution. From Fig. 7.4, the dimensions are,

$$A = (7{\cdot}2 - 2 \times 3{\cdot}32)/\ 2 = 0{\cdot}28 \text{ m} = 280 \text{ mm}$$

$$L = 7200 \text{ mm}, H = 250 \text{ mm}, R = \frac{1200}{2} = 600 \text{ mm.}$$

Load on each saddle support is,

$$R_a = R_b = \frac{103 \cdot 2}{2} = 51{\cdot}6 \text{ kN}$$

Bending moment at supports from equation (7.25) is,

$$M_a = 51{\cdot}6 \times \frac{280}{1000} \times \left\{1 - \frac{1-(280/7200)+[(600^2-250^2)/(2\times 280\times 7200)]}{1+(1{\cdot}33\times 250/7200)}\right\}$$

$$= 0{\cdot}156 \text{ kNm}$$

also bending moment at mid-span from equation (7.26) is,

$$M_c = \frac{51{\cdot}6\times 7200}{4\times 1000}\left\{\frac{1+[2(600^2-250^2)/(7200)^2}{1+(1{\cdot}33\times 250/7200)} - \frac{4\times 280}{7200}\right\}$$

$$= 75{\cdot}35 \text{ kNm}$$

Using equation (7.27) and (7.28), the stresses in the topmost and bottommost fibres will be,

$$f_{to} = f_{bo} = \frac{0{\cdot}156(10)^6}{(600)^2\pi(10-2)\times 1{\cdot}0} = 0{\cdot}017 \text{ N/mm}^2$$

Stress in the shell at mid-span from equation (7.29) is,

$$f_{ms} = \frac{75{\cdot}35(10)^6}{(600)^2\pi(10-2)} = 8{\cdot}33 \text{ N/mm}^2$$

Axial stress in the shell wall due to internal pressure using equation (3.2) is,

$$f_a = \frac{0{\cdot}45\times 1200}{(4\times 10)} = 13{\cdot}5 \text{ N/mm}^2$$

The combined stresses are,

$$f_a + f_{to} = 13{\cdot}5 + 0{\cdot}017 = 13{\cdot}517 \text{ N/mm}^2$$

$$f_a - f_{bo} = 13{\cdot}5 - 0{\cdot}017 = 13{\cdot}483 \text{ N/mm}^2$$

$$f_a + f_{ms} = 13{\cdot}5 + 8{\cdot}33 = 21{\cdot}83 \text{ N/mm}^2$$

All these combined stresses are well within . .mit of 92 N/mm^2, hence, the provided saddle supports are quite sufficient.

Nomenclature

A : distance between saddle support center to the end of cylindrical vessel, m

A_b : cross section area of bolts, mm^2

A_c : cross section area of column, mm^2

A_e : effective cross section area of bolts, mm^2

A_f : area of contact between bearing plate and foundation concrete surface, m^2

A_r : effective cross section area of stiffener ring, mm^2

b : width of the gusset plate, mm
b_a : width of rolled steel angle, mm
b_r : circumferential width of bearing plate, mm
c : width of the stiffener or base plate, mm
D_o : outer diameter of vessel, mm
D_{sk} : outer diameter of cylindrical skirt, mm
D_p : pitch or bolt circle diameter, m
d_B : outer diameter of bearing plate, m
e : eccentricity of load in column from vessel surface, mm
f : maximum combined direct and bending stress, N/mm^2
f_a : maximum stress in base plate, N/mm^2
f_{ac} : allowable compressive stress in stiffener ring material, N/mm^2
f_{bo} : maximum stress in bottom most fiber of shell, N/mm^2
f_{bolt} : stress in bolt, N/mm^2
f_c : allowable compressive stress in column, N/mm^2
f_{cb} : minimum stress between bearing plate and foundation concrete surface, N/mm^2
f_{ce} : bending stress due to eccentricity of load, N/mm^2
f_{cs} : maximum compressive stress in skirt, N/mm^2
f_d : stress in skirt due to dead load, N/mm^2
f_e : maximum compressive stress in gusset plate edge, N/mm^2
f_{max} : maximum stress in bearing plate elemental strip of unit width, N/mm^2
f_{ms} : stress in shell wall at saddle support mid-span, N/mm^2
f_t : allowable flexural stress in column, N/mm^2
f_{te} : maximum tensile stress in skirt wall, N/mm^2
f_{to} : maximum stress in top-most fiber of shell, N/mm^2
f_w : bending stress in due to wind load in lug supports, N/mm^2
f_{wb} : bending stress in due to wind load at vessel base on skirt supports, N/mm^2
H : height of dished head, mm
H_1 : height of vessel top above foundation level on lug supports, m
H_2 : height of vessel bottom above foundation level on lug supports, m
h : height of gusset plate, mm
h_s : height of vessel shell above base on skirt supports, m
h_1 : height of lower part of on skirt supports, m
h_2 : height of upper part of on skirt supports, m
K : coefficient of shape factor

K', K'' : coefficient as a function of saddle angle for stiffener ring
K_s : coefficient of stability
$K_{s(min)}$: minimum value of coefficient of stability
$k_1,\ k_2$: coefficient as a function of saddle angle for shell wall
L : length of vessel shell, m
L_c : length of the column, mm
l_e : effective length of column, mm
l_r : distance between skirt and outer radii of bearing plate, mm
M : bending moment for each gusset plate, kNm
M_a : bending moment at saddle support, kNm
M_c : bending moment at span center between saddles, kNm
M_d : bending moment due to eccentric self weight of vessel in skirt supports, kNm
M_{lw} : bending moment at the base due to wind in lower part of vessel, kNm
M_{uw} : bending moment at the base due to wind in upper part of vessel, kNm
M_w : total bending moment at the base due to wind load on vessel, kNm
M_{max} : maximum bending moment in bearing plate, Nmm
N : number of bolts for skirt bearing plate
n_b : number of bracket supports
n_c : number of columns
P : maximum load on base plate, N
P_{bolt} : load on each bolt in skirt supports, N
P_{lw} : wind force acting on lower region of skirt supports, N
P_{uw} : wind force acting on upper region of skirt supports, N
P_w : wind force acting on vessel on lug supports, N
p : pressure due to total loads on vessel, N/mm^2
p_a : average pressure on base plate, N/mm^2
p_1 : wind force acting on lower region of skirt supports, N/mm^2
p_2 : wind force acting on upper region of skirt supports, N/mm^2
R : radius of vessel shell, mm
R_a : reaction at saddle support *a*, kN
R_b : reaction at saddle support *b*, kN
R_m : moment arm in saddle support, m
r_y : radius of gyration of stiffener ring, mm
t_a : thickness of rolled steel angle, mm
t_B : thickness of bearing plate, mm
t_g : thickness of gusset plate, mm

t_s : thickness of shell wall, mm
t_{sk} : thickness of cylindrical skirt, mm
t_{st} : thickness of stiffener plate in saddle, mm
W : total dead load including self weight of vessel, its contents, attachments and insulating material etc, N
W_w : total load due to wind on vessel skirt supports, N
W_1 : load due to self weight of vessel only, N
W_2 : load from vessel contents, N
w : uniformly distributed load of vessel on saddle supports, kN/m
Z_p : section modulus of gusset plate, mm^3
Z_y : section modulus of stiffener ring, mm^3
α : angle of gusset plate edge with horizontal, deg
θ : saddle angle, deg

8

Grain Dryers

8.1 Introduction

Drying of grain or food product implies the removal of water from it by vaporization. It is defined as the universal method of conditioning the material by removing moisture to such a level that it is in equilibrium with the normal atmospheric air in order to preserve the quality and nutritive value of the product and seed quality. The term conditioning refers to maintain uniform moisture and temperature throughout the dried product by moving an air through it. The dried product can be stored for a longer period without any deterioration. It is due to the fact that micro-organisms responsible for the product deterioration, spoilage and decay are unable to grow and multiply in absence of sufficient moisture. Moreover, many enzymes promoting undesired changes in the product cannot function properly without water. A product is in equilibrium with its surrounding environment when the rate of moisture loss from the product to the surrounding atmosphere is equal to the rate of moisture gain of the product from the surrounding atmosphere. The moisture content of the product when it is in equilibrium with the surrounding atmosphere, is called as the equilibrium moisture content or hygroscopic equilibrium. Thermodynamically, the equilibrium is reached when the free energy change for the product is zero. The moisture content of the solid is usually expressed as the moisture content by weight of bone dry material in the solid.

When the product is exposed to heat during drying, simultaneously, two processes are involved : (*a*) the transfer of heat energy from the surrounding environment to evaporate the surface moisture, and (*b*) transfer of internal moisture of the product to its surface with the subsequent evaporation. The rate of drying is governed by the rate at which the two processes proceed. Heat energy from the surrounding (air) to the surface of the product occurs through conduction, convection or radiation or combinations of the three modes. In most of the dryers heat is supplied to the product surface and thence to the interior except in some cases, such as in dielectric or microwave drying where the energy is supplied to generate heat internally within the product and flows to the exterior surfaces. The rate of water removal in the form of vapour from the product surface depends on the external conditions of temperature, pressure, air flow rate, air humidity and area of exposed surface. The internal movement of water within the product is a function of the physical nature of the solid, its moisture content and temperature.

8.2 External conditions for drying

Psychrometric chart gives the conditions of the surrounding of grain to be dried. It is a graphical presentation of the relationship between the air and water vapour associated with it on two main axes of the absolute humidity (kg/kg) as ordinates and dry bulb temperature (°C) as abscissa (Fig. 8.1). Thus, every point (coordinates) on psychrometric

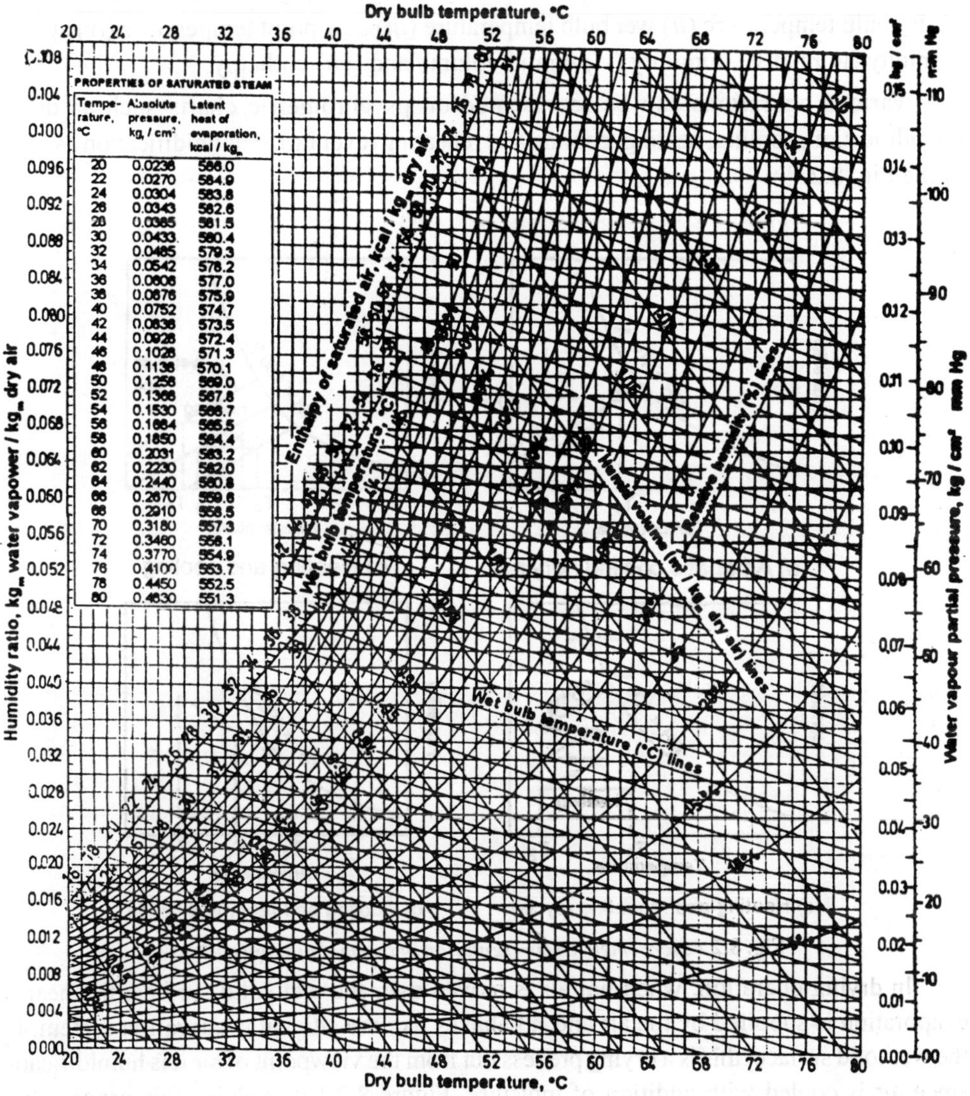

Temperature, °C	Absolute pressure, kg_f / cm²	Latent heat of evaporation, kcal / kg_m
20	0.0238	586.0
22	0.0270	584.9
24	0.0304	583.8
26	0.0343	582.6
28	0.0385	581.5
30	0.0433	580.4
32	0.0485	579.3
34	0.0542	578.2
36	0.0606	577.0
38	0.0676	575.9
40	0.0752	574.7
42	0.0838	573.5
44	0.0928	572.4
46	0.1026	571.3
48	0.1136	570.1
50	0.1258	569.0
52	0.1368	567.8
54	0.1530	566.7
56	0.1664	565.5
58	0.1850	564.4
60	0.2031	563.2
62	0.2230	562.0
64	0.2440	560.8
66	0.2670	559.6
68	0.2910	558.5
70	0.3180	557.3
72	0.3460	556.1
74	0.3770	554.9
76	0.4100	553.7
78	0.4450	552.5
80	0.4830	551.3

Fig. 8.1 *Psychrometric chart*

chart represents the condition of air in respect of humidity and temperature. The saturation curve (100% RH) is plotted on left side with the dry bulb temperature on x-axis (zero per cent RH). The whole unsaturated region is divided evenly by different levels of constant relative humidity curves. Selecting a point on saturation curve a straight line with slope ($-\lambda C_s$) running down into the unsaturated region which is a wet bulb line also called as the adiabatic cooling line. Humid volume lines steeper than wet bulb lines are also plotted in the chart along which the volume of humid air is constant and are not perfectly parallel to each other. The numerical values of the enthalpy lines are given on the scale beyond the saturation line. Knowing any two values of the properties of admixture of air and water vapour, the other properties can be easily determined by using the psychrometric chart, namely, (*i*) dry bulb temperature (*ii*) wet bulb temperature (*iii*) dew point temperature (*iv*) absolute humidity (*v*) relative humidity (*vi*) humid volume and (*vii*) enthalpy.

Various operations can be represented on psychrometric charts such as drying, humidification, cooling, heating, cooling and reheating (cycling), dehumidification as shown in the Fig. 8.2.

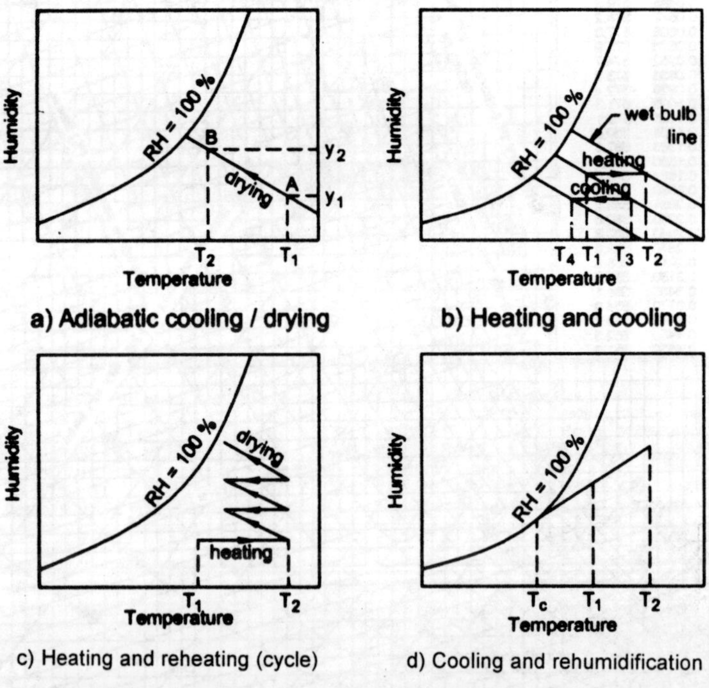

Fig. 8.2 *Psychrometric representation of thermal unit operations*

In drying operation, when hot air is moved over the wet solid surface, the heat for evaporation is supplied by the air and it behaves like the adiabatic saturation system. For the wet solid surface, this is a drying process but from the viewpoint of air it is humidification since air is cooled with addition of moisture. Figure 8.2 (*a*) explains this process from point A (t_1, Y_1) to point B (t_2, Y_2) along the wet-bulb line.

Air cooling takes place along horizontal constant humidity line. [Fig. 8.2 (*b*)] until it reaches the saturation line at its dew point. Heating is in opposite direction of cooling along horizontal line increasing the dry-bulb temperature.

In many dryers intermittent heating is provided for some desirable properties of the product. In this case the air is reheated in order to reduce its relative humidity and thus, providing an additional capacity to air to evaporate more moisture from the solid surface being dried. Figure 8.2 (*c*) shows the alternate heating, drying and reheating in a cyclic manner.

8.3 Internal conditions for drying

When the moisture is evaporated from the surface of the grain or solid, the temperature gradient is developed within it, due to heat transfer resulting into moisture migration from solid body to its surface. This is mainly due to the diffusion, capillary movement, internal pressure during drying set up by shrinkage of the solid or by repeated vaporization and condensation of moisture on exposed surface in case of indirect drying process. The quality of air, its residence time and temperature gradient set-up are the variables affecting the drying rate and time.

Moisture content

Rockland (1969) expressed three types of water in grain or food products *viz*., (*a*) water molecules containing inside the cell walls of the plant structure which are chemically bound present as liquid solution and exerts vapour pressure less than that of the pure liquid, is called bound water, (*b*) unbound water in excess of bound water present in interstitial pores which exerts vapour pressure equal to that of the pure liquid at the same temperature, (*c*) free water present in the product (solid) depends upon the vapour concentration in air, in excess of equilibrium moisture. It is most difficult to remove bound water and easier for unbound water and free water in order. Only free water can be evaporated during drying.

The vapour pressure at equilibrium around the vicinity of the interface between free water and vapour is the saturated vapour pressure which exists in very moist products. When the moisture is removed, the dissolved solids concentration is increased which reduces the vapour pressure due to osmotic forces. Further removal of water results in remainder of the bound water only within the product considerably lowering the vapour pressure. Such product can only be in equilibrium with the external atmospheric conditions.

Equilibrium moisture content

Equilibrium moisture content occurs under a given vapour pressure of water in the surrounding air which is dependent on the temperature, water content of the product, and its nature and the soluble solids present in it. Usually, the water activity of the product at its surface is considered. It is defined as the ratio of the partial pressure of water in the product to the vapour pressure of water at the same temperature. The equilibrium moisture content determines whether the grain or product will lose or gain the moisture under the given conditions of temperature and relative humidity, thus, directly concerned with the drying operation. In case of grain the equilibrium moisture content depends upon the variety, type and its maturity. A sigmoid (S-shape) curve of the equilibrium relative humidity and the moisture content of the material at a particular temperature is known as equilibrium moisture curve or isotherm.

8.4 Heat transfer in drying

Drying of the product is a thermal process and several products can be dried only by heating them above the boiling point of the liquid. In most of the adiabatic drying processes, diffusion of vapour through solids is present and it makes the process complicated. Most of the dryers are designed on the consideration of heat transfer only. In drying, heat is required for (*a*) heating the product to the vaporization temperature, (*b*) vaporization of the water, (*c*) heating the product to its final temperature and (*d*) heating the vapour to its final temperature. Drying may be accomplished by all the three modes of heat transfer : conduction, convection and radiation, usually one of the mechanism predominates to govern the heat transfer depending on the type of drying equipment used. For the drying operation of solids, major quantity of heat is required for vaporization of water, the other requirements may be negligible. The quantity of heat transferred per unit mass in the dryer is given by the following equation.

$$\frac{Q}{m_p} = C_{pp}(T_{p2} - T_{p1}) + X_1 C_{pw}(T_v - T_{p1}) + (X_1 - X_2)\lambda_w + X_2 C_{pw}(T_{p2} - T_{v1}) + (X_1 - X_2) C_{pv}(T_{v2} - T_{v1}) \tag{8.1}$$

Hence, the specific heats and vaporization heat are assumed to be constant. Further, it is also assumed that the vaporization of moisture occurs at a constant temperature T_v. Although these conditions seldom occur, equation (8.1) provides approximately good estimates of heat transfer rates.

In case of dryers working with adiabatic heating, change in enthalpy can be determined from psychrometric chart or by using following equation (8.2),

$$Q = m_g(1 + H_1) C_{s1}(T_{g1} - T_{g2}) \tag{8.2}$$

8.5 Mass transfer in drying

Mass transfer takes place under the influence of a driving force provided by the difference in the mass concentration or partial pressure and the rate is directly proportional to it and the system properties characterized by the mass transfer coefficient. The average mass transfer rate can be computed by using the following equation,

$$m_v = m_p(X_1 - X_2) \tag{8.3}$$

However, application of equation (8.3) is not simple as the equation for rate of heat transfer, mainly due to the change in the moisture content with the progress of drying. Firstly, the moisture is evaporated from the product surface and then from deeper within the product to the surface and thence, to the air. Thus, firstly, the relation between the moist surface of the product and drying air should be established and then the moisture diffusion through the product would be considered. The heat and mass transfer should be simultaneously considered for such study.

8.6 Drying mechanism

When the solids like grain are dried with hot air, the rate of drying (rate of water removal) is dependent on the drying air conditions, nature of the solid to be dried and other factors related to the dryer parameters. The plot of moisture content of the product versus drying time (Fig. 8.3) shows different drying periods. It is obvious that free water on the saturated surface of the product will be evaporated first, followed by moisture from the interior of the product.

8.6.1 Constant rate period

The rate of moisture removal under constant rate drying period is independent of the moisture content in the product. It is just the evaporation of water from the product surface or within it under any drying air conditions. This period continues till the interior moisture is migrated more rapidly than the evaporation rate on the product surface.

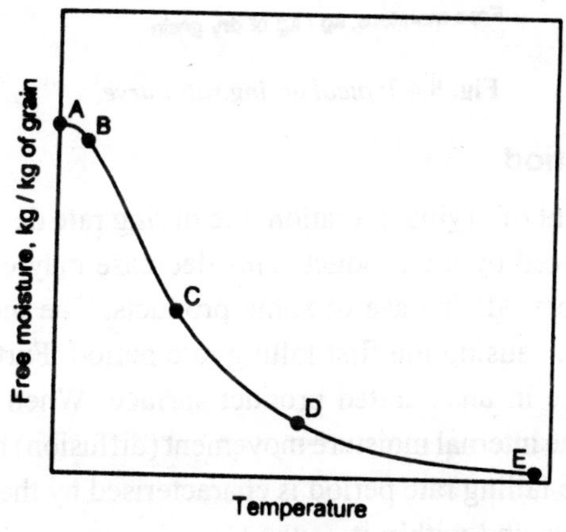

Fig. 8.3 *Drying rate curve for grain*

The moisture removal rate can be related to the heat transfer rate at constant product temperature since, all the heat supplied is used completely for water evaporation. Thus, for the same set of conditions both these factors must yield the same rate of moisture removal or drying rate which is expressed below,

$$\frac{dm_v}{dt} = (hA_w / \lambda_w)(T_a - T_w) = K_w A_w (H_w - H_a) = K_w A_w (p_w - p_a) \tag{8.4}$$

With the constant drying conditions the product saturated surface attains the wet bulb temperature in adiabatic drying process. Drying takes place at the product surface by water evaporation and the moisture migrates by a driving force of vapour pressure difference. As the product moisture content decreases, the constant rate period ends at certain moisture content and with further drying, the drying rate decreases. This moisture content is known as the critical moisture content as shown in Fig. 8.4. It varies with the product bed thickness, drying rate, drying conditions, surface moisture concentration and product characteristics such as the shape and size.

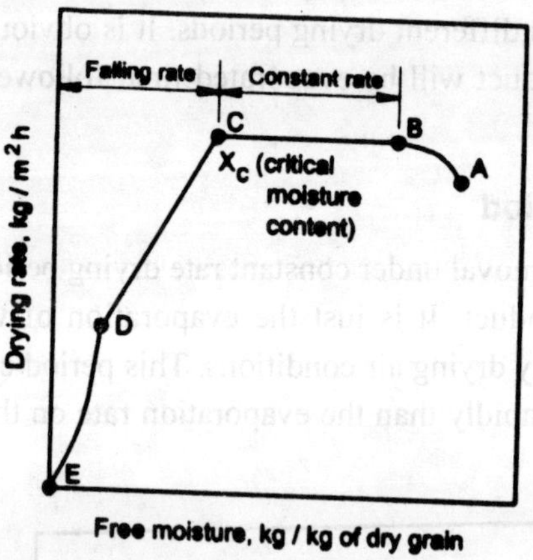

Fig. 8.4 *Typical drying rate curve*

8.6.2 Falling rate period

With the advancement of drying operation, the drying rate decreases after the critical moisture content is achieved by the product. This decrease may either be linear or there may be two falling rate periods in case of some products. The moisture conveyed from within the product is less, causing the first falling rate period. Further decrease in drying rate is due to the increase in unsaturated product surface. When the product surface is completely unsaturated, the internal moisture movement (diffusion) becomes the controlling factor for drying rate. The falling rate period is characterised by the increased temperature both at the product surface and within it.

8.6.3 Estimation of drying time

Drying rates and drying time are the most essential and important quantities under the conditions provided in dryers, which determine the capacity of it. In most of the dryers, drying rates vary with the time and product moisture content. Under such situation, the drying time can be assessed from the experimental drying curve (Fig. 8.3) drawn under constant drying conditions namely, product surface area for drying, relative humidity and temperature. A batch of moist material is dried and change in product weight is continuously recorded. This curve can then be transformed into a drying rate curve (Fig. 8.4). Under constant rate period, the drying rate can be written as,

$$R_c = -\frac{m_p}{A} \cdot \frac{dX}{dt} \tag{8.5}$$

Integrating equation (8.5) for constant rate period within the initial and final moisture content , the following expression can be obtained for drying time,

$$t_c = \frac{m_p}{A} \cdot \frac{(X_1 - X_2)}{R_c} \tag{8.6}$$

Under falling rate period, from equation (8.5), the drying time can be determined by using the following equation when $X_1 < X_c$,

$$t_f = \frac{m_p}{A} \int_{X_2}^{X_1} \frac{1}{R_f} \cdot dX \tag{8.7}$$

A curve for $1/R_f$ versus X drawn from the experimental data can be used to determine the drying time under falling rate period.

8.6.4 Diffusion-controlled falling rate period

After the product surface is unsaturated with water, the moisture from inside the product is transferred through the solid by diffusion. It controls the drying rate and the moisture content at the product surface is at or very near to the equilibrium value. Air velocity has little or no effect and the relative humidity influences the drying process primarily through its effect on equilibrium moisture content. Since, the diffusivity increases with an increase in temperature, the drying rate increases with the product temperature. Equation for diffusion for long drying time and spherical particles is,

$$\frac{X_1 - X}{X_1 - X_e} = 1 - \frac{6}{\pi^2} \cdot \exp\left(\frac{-D\pi^2 t}{R^2}\right) \tag{8.8}$$

Equation (8.8) can be solved for the drying time to yield,

$$t_f = \frac{R^2}{D\pi^2} \ln\left[\frac{X_1 - X_e}{X - X_e} \cdot \frac{6}{\pi^2}\right] \tag{8.9}$$

8.6.5 Characteristic drying curve

Most of the drying rate curves appear geometrically similar for drying of solids and are the function of extent to which drying has occurred. For a specific material, if these curves are normalized with respect to an average moisture content and initial drying rate all result in a single curve called the characteristic drying curve (Fig. 8.4) and the characteristic drying rate and characteristic moisture ratop are defined as,

$$Rch = \frac{N_v}{N_w} \tag{8.10}$$

and $$\phi_{ch} = \frac{X - X_e}{X_c - X_e} \tag{8.11}$$

It is necessary for the characteristic drying curve to satisfy the two conditions, namely (*i*) the critical moisture content is unchanged and independent of the initial moisture content and external conditions, and (*ii*) for a specific solid material all the drying curves show unique shape and independent of external conditions. This occurs in the product of the same form.

8.7 Classification of dryers

The dryers can be classified in four groups based on the mode of heat transfer *i.e.* conduction heating, convection heating, radiant heating and dielectric heating. Freeze drying is a special case of conduction heating. The second way of classifying the dryers on the basis of type of vessel used for drying *i.e.* tray dryer, rotating drum dryer, spray dryer, pneumatic dryer or fluidized bed dryer. The operational mode can also be used for classifying the dryers *e.g.* continuous, semicontinuous and batch type. In most of the grain dryers, heat transfer occurs through convection.

8.8 Factors affecting the drying process

The rate of product drying is a function of drying air temperature and flow rate, drying time, relative humidity, initial moisture content, depth of product in the dryer, type and size of the product particles etc. The effect of some of these factors are discussed below.

(1) *Air temperature* : The drying rate of the product is directly proportional to the difference in dry bulb and wet bulb temperatures (equation 8.4). Thus, the drying rate is faster for the higher drying air temperatures for drying the grain, the equilibrium moisture content decreases with an increase in drying air temperature. It is necessary to adopt a particular range of temperature depending upon the type of product to be dried. Protein denaturation and coagulation may take place in some grains for temperature above 50 °C.

In presence of high moisture and temperature above 70 °C, gelatinization and partial conversion into dextrin takes place resulting into deterioration of the product colour. An enzymic activity on fats increases with the increase in moisture and temperature in the range 40-50 °C, but it decreases with further temperature rise and is completely ceased in the range of 80-100 °C. At high temperature the heat sensitive vitamin B is destroyed from germ and aleurone layer of the seed. In addition seed may also lose the viability. For seed, food and feed, the recommended temperature ranges are 40-45, 40-60 and 80-85 °C, respectively.

(2) *Air flow rate* : The internal resistance to moisture movement through the granular product is higher than the resistance for surface mass transfer. Thus, the air flow rate past the product particles has no significant effect on the drying time and the drying coefficient (10-68 $cm^3/s/cm^2$) but higher air flow rates have some effect on drying rate.

(3) *Drying time* : Drying rate of the product is proportional to the drying time (Fig. 8.3). With the advancement in drying time, the drying rate is reduced due to reduction in product moisture content. For intermittent drying, the exposure time in each pass affects the drying rate. A sum of all the exposure times in each pass gives the total drying time.

(4) *Relative humidity* : The relative humidity of the drying air has lesser effect as compared with that of the temperature changes. The drying rate decreases slightly with the increase in the drying air humidity because of moisture involved from the air side.

(5) *Initial moisture content of product* : It decides the quantity of moisture to be handled by the drying air. Higher the initial moisture, the drying rates are also higher since more amount of free moisture is available for evaporation.

(6) *Depth of product in dryer* : The grain can be dried either by thin layer or deep bed drying. The former is discussed above. The drying air condition is not similar through the product mass but varies with the drying time and depth of product in dryer. Figure 8.5 shows the deep bed drying conditions. The total bed can be considered as several thin iayers of the product. The temperature and humidity of the drying air entering and leaving each layer depends upon the drying time, drying stage and the moisture removed from dried layer up to X_e. A small quantity of moisture is added to the upper zone above drying front till the later reaches to it. The time of advance of drying front to reach the top layer is expressed, as,

$$t_{d1} = \frac{W_d}{100\, A_d\, m'_a} \cdot \frac{X_1 - X_a}{H_s - H_1} \tag{8.12}$$

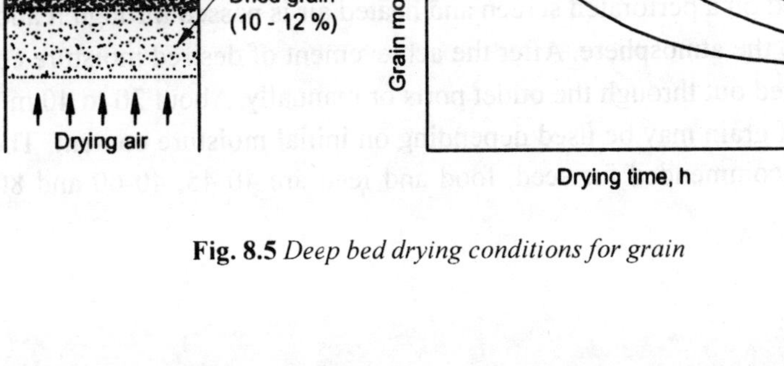

Fig. 8.5 *Deep bed drying conditions for grain*

As soon as the drying front reaches the top, a decreasing rate period starts, where the drying time is given by the expression,

$$t_{d2} = \frac{1}{K_d} \ln \left[\frac{X_a - X_e}{X - X_e} \right] \tag{8.13}$$

In deep bed drying, the bottom layer is dried more than the top layer developing moisture gradient within the bed and leading to non-uniform drying of product. This does not allow to use higher temperatures for deep bed drying. Usually, 40 °C temperature or 15 °C temperature rise can be used.

8.9 Drying equipment

Grain dryers are of two types, unheated air dryers and heated air dryers. The former type circulates ambient air through the grain bed in order to aerate and remove little moisture from the grain. Heated air dryers use the air heated above ambient temperature, the rise being from 15 to 60 °C depending on the type of grain to be dried and their heat sensitiveness as discussed earlier. Heated air dryers are of two types, namely, flat bed batch dryers and continuous flow or recirculatory dryers ; according to the state of drying product in the dryers.

8.9.1 Flat bed batch dryer

This is the most commonly used heated air dryer which is stationary using deep bed of grain in batches. It has a simple design and operation, can be manufactured locally with common materials like steel sheet, angles or wood etc. It can be used for drying of seed and also for post-drying period. It is low cost and is mostly used in many countries for on-farm drying. However, this dryer has limitations of slow drying rate and low holding capacity as compared with the continuous flow dryer resulting in uneven drying.

Batch dryer consists of a rectangular drying chamber below which plenum chamber is provided. Both are divided with the help of a perforated bottom over which the grain is fed. It may be horizontal or inclined (Fig. 8.6) through which hot air is passed and distributed through the grain mass. An electric heater, oil-fired burner or waste fired furnace with heat exchanger is used as a source of heating the air. Atmospheric air is forced by the fan (blower) with 0·25 to 1·0 horse power electric motor. All the components are mounted on frame resting on ground or suitable foundation.

Grain is loaded on a perforated screen and heated air is passed through it and exits at the grain surface to the atmosphere. After the achievement of desired moisture content in grain, it is discharged out through the outlet ports or manually. About 20 to 40 m^3/min air flow rate per ton of grain may be used depending on initial moisture content. The drying air temperatures recommended for seed, food and feed are 40-45, 40-60 and 80-85 °C, respectively.

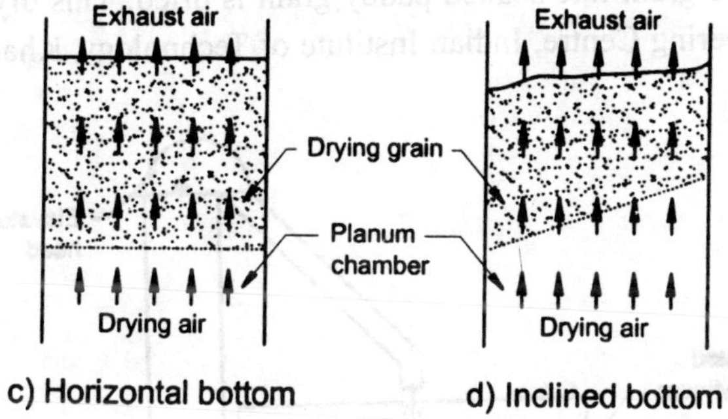

Fig. 8.6 *Batch dryer aeration system*

8.9.2 Continuous flow dryer

In case of continuous flow or recirculatory dryers, the grain is flowing continuously by controlled rate under gravity either through a concentric path (RPEC dryer) or in zigzag path (LSU and baffle dryers). While flowing, the grain may be mixed due to free flowing conditions (LSU baffle dryers) or it may be non-mixing due to restricted path (RPEC dryer). Accordingly, they are called as mixing or non-mixing type dryers, respectively.

8.9.2.1 RPEC dryer

It consists of two concentric perforated (2 mm diameter) mild steel sheet cylinders mounted 150 to 200 mm apart creating a grain flow path (Fig 8.7).

A hopper bottom is provided to the cylinders for easy discharge of grain under gravity. Inner cylinder is also provided with a cone at the top for even distribution of grain around the periphery of the concentric path. A complete structure is elevated suitably and mounted on frame made out of steel sections. Grain is discharged through the inclined pipe to the boot of a bucket elevator which feeds the grain again to the top of the cylinders. Hot air supplied by a centrifugal blower is forced to the inside of the inner cylinder (plenum chamber). The heat source may be as mentioned above. Hot air is forced through the perforations and grain flowing downward and exit from the outer cylinder. Thus, the path of hot air is at right angle to the grain flow path.

Initially, the grain is fed to the hopper provided on ground level to the boot of the bucket elevator and feed to the top cone. While flowing down under gravity the hot air passes across it and drying is achieved. Grain flow rate can be controlled with the help of a flap valve provided in the discharge pipe. Grain can be recirculated till it is dried to the desired moisture content. This dryer is simple, easy operation and maintenance, low-cost and can be used on-farm. However, the drying is not uniform when compared with mixing type dryers and the perforations are required to be cleaned frequently when fibrous high moisture grain like soaked paddy grain is dried. This dryer is developed at Rice Process Engineering Centre, Indian Institute of Technology, Kharagpur.

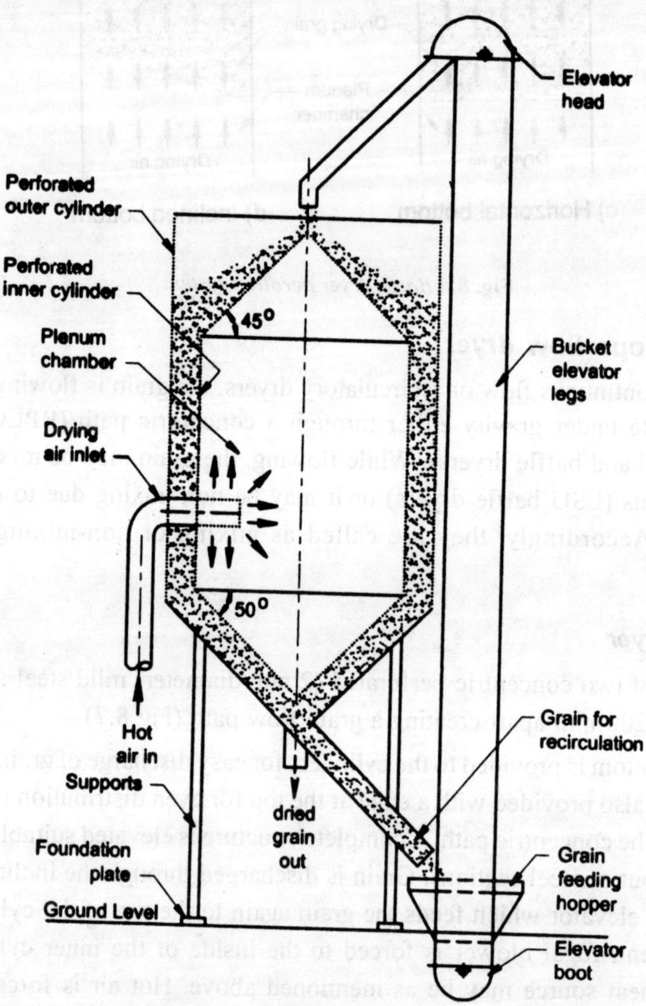

Fig. 8.7 *RPEC recirculatory batch dryer*

8.9.2.2 LSU dryer

This continuous mixing type dryer is developed at the Lousiana State University, USA and is widely used for uniform grain drying. However, its cost is higher than the other dryers. It consists of a square chamber divided into two parts *i.e.*, holding bin at top and the drying chamber at bottom (Fig. 8.8). The drying chamber consists of inverted V ports arranged in a horizontal row, but zigzag in vertical position *i.e.*, each successive layer is offset by half centre to centre distance so as to split the grain stream in two parts, consequently mixing with the other side stream at each layer. Ribbed rollers are provided at the bottom of the drying chamber rotating at low speed and controlling the grain feed rate by adjustment of the roller speed. Grain discharged through the rollers, is collected in the hopper bottom below and out either for recirculation or out of the dryer.

Air heated by any source as mentioned above is circulated through the drying chamber. For this purpose one end of each inverted V ports is open and other is closed. Thus, alternate layer ports are air inlets and outlets. Air is fed to the alternate layer ports only, so also the exhaust through alternate layer ports as shown in Fig. 8.8. All the ports have 90° angle at top, equispaced ; the number depending on the dryer size. Air is forced through one side of the dryer through many ports and exhausted at the other end through alternate ports. The bottom of port is open for air distribution in the drying chamber.

Grain is fed to the top of holding bin and successively to the drying chamber through a subdividing and mixing zigzag path. The hot air entering through the ports comes directly in contact with the grain and drying is achieved. Further, grain passes through the feed rollers to hopper bottom and out of the dryer. The air after heating the grain, is exhausted through the outlet ports.

This type of dryer has the capacity of about 2 to 12 t grain or even higher capacities can be provided. Air flow rate of about 60 to 70 m^3/min/t of grain may be provided. The power requirement will vary according to the dryer.

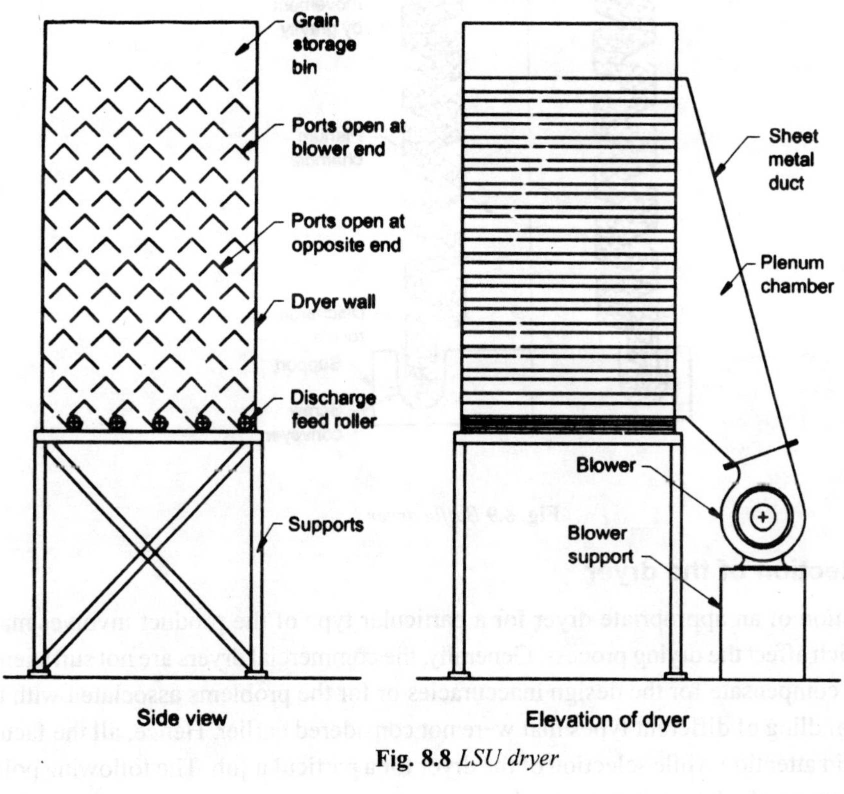

Fig. 8.8 *LSU dryer*

8.9.2.3 Baffle dryer

It is a continuous flow mixing type of grain dryer giving a uniformly dried product. However, it has the high ratio of plenum volume to the total volume of the dryer and the grain movement on baffles is slow. The baffle dryer consists of a grain inlet chamber, drying chamber with baffles, plenum chamber, grain feed rollers and hopper bottom (Fig. 8.9). The drying chamber is made of a vertical column with the zigzag grain path by providing a number of baffle plates where grain mixing takes place to certain extent. The two baffle plates are fitted about 200 mm apart, one on inner side and the other on outer since. The grain feed rollers and hopper bottom are similar to the LSU dryer.

Grain is fed to the inlet chamber and moves through a prescribed path through drying chamber where it encounters a cross flow of hot air from plenum chamber. Partially dried grain is discharged from the bottom of hopper and may be recirculated for further drying.

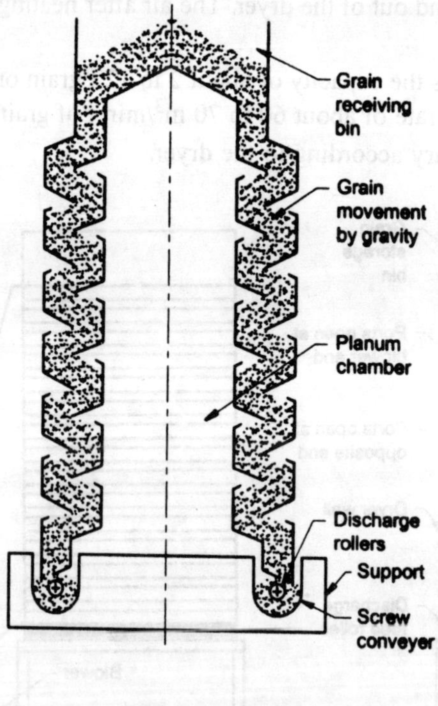

Fig. 8.9 *Baffle dryer*

8.10 Selection of the dryer

Selection of an appropriate dryer for a particular type of the product involves many factors which affect the drying process. Generally, the commercial dryers are not sufficiently flexible to compensate for the design inaccuracies or for the problems associated with the material handling of different types that were not considered earlier. Hence, all the factors shall be paid attention while selection of the dryer for a particular job. The following points should be attended while selecting the dryer.

1. Properties of the product should be considered.

2. Product drying characteristics such as drying curves, drying time, nature of moisture with the product, equilibrium, initial and targeted moisture content should be obtained.

3. Type of material handling during drying operation *i.e.* quantity, batch or continuous flow, method of feeding and discharge, flow control mechanism, intermittent drying operation etc.

4. Quality of the end product in terms of uniformity of drying, product temperature, cracking of the product particles, colour and flavour of the product should be observed.

5. Availability of heat source, suitable ambient conditions, space for installation at site should also be considered.

6. The dryer should be simple in design, operation and maintenance, low-cost, multipurpose and portable, if required. It should have a proper capacity to match the demand of the users.

7. Techno-economic feasibility of the dryer should be worked out for its adoption in the field.

8.11 Dryer calculations

The heat capacity or heat supplied by the drying air is given by the following expression.

$$q_a = (0{\cdot}24 + 0{\cdot}45\ H_0)\ m_a\ (T_{a1} - T_{a2})\ t \tag{8.14}$$

The quantity of heat required for drying of the product can be determined as below.

(*a*) sensible heat of the product

$$q_1 = W_d\ C_{pp}\ (T_{p2} - T_{p1}) \tag{8.15}$$

(*b*) sensible heat of water

$$q_2 = W_d\ X_1\ C_{pw}\ (T_{p2} - T_{p1}) \tag{8.16}$$

(*c*) latent heat of water vaporization

$$q_3 = W_d\ (X_1 - X_2)\ \lambda_w \tag{8.17}$$

Total heat required for product drying is then,

$$q_d = q_1 + q_2 + q_3 + q'$$

where, q' is generally assumed to be about 10% of the total heat required for drying of the product.

Then
$$\begin{aligned} q_d &= 1{\cdot}1\ (q_1 + q_2 + q_3) \\ &= 1{\cdot}1\ [W_d\ C_{pp}\ (T_{p2} - T_{p1}) + W_d X_1\ C_{pw}\ (T_{p2} - T_{p1}) + W_d\ (X_1 - X_2)\lambda_w] \\ &= 1{\cdot}1\ W_d\ [(C_{pp} + X_1\ C_{pw})\ (T_{p2} - T_{p1}) + (X_1 - X_2)\ \lambda_w] \end{aligned} \tag{8.18}$$

Now, the heat supplied by the hot air is equal to the heat utilized in the dryer that is,

heat supplied = heat consumed in the dryer

or
$$(0{\cdot}24 + 0{\cdot}45\ H_0)\ m_a(T_{a1} - T_{a2})\ t = 1{\cdot}1\ W_d[(C_{pp} + X_1\ C_{pw})\ (T_{p2} - T_{p1}) + (X_1 - X_2)\ \lambda_w] \tag{8.19}$$

then, $$m_a = \frac{1 \cdot 1 W_d [(C_{pp} + X_1 C_{pw})(T_{p2} - T_{p1}) + (X_1 - X_2)\lambda_w]}{(0 \cdot 24 + 0 \cdot 45 H_0)(T_{a1} - T_{a2})\, t} \quad (8.20)$$

This, mass flow rate can be converted into volumetric air flow rate by multiplying with humid volume for the given conditions from the psychrometric chart, that is,

$$V_a = m_a V_h \quad (8.21)$$

Determine the power requirement for blower as below.

Compute the air passage area and the air flow requirement per unit area of air passage. From the Shedd's curve (Agricultural Engineering Handbook) find out the pressure drop for air flow and bed thickness of grain. If given thickness is not readable, interpolate to get the pressure drop. Determine the power requirement using the following expression.

$$P = \frac{\Delta p\, m_a \cdot g}{59176} \quad (8.22)$$

Example 8.1 : *The parboiled paddy at 33% (wb) is to be dried to 16% (wb) moisture content for which the angle of repose is 44° and the bulk density at 16% mc wb is 580 kg/ m^3. The product at an ambient temperature is fed to the RPEC dryer leaving at 70 °C. The ambient air is at 28 °C and 68% relative humidity. The heated drying air enters the plenum chamber at 87 °C and exit at 42 °C. Assuming the total drying time of 3.5 h in parboiling process, design the dryer for 2·5 t holding capacity with 200 mm grain thickness.*

Solution. Referring to Fig. 8.10, assume the plenum chamber diameter $D_2 = 1 \cdot 30$ m, grain outlet diameter, $D_3 = 0 \cdot 15$ m, then,

$$D_1 = 1 \cdot 30 + 2 \times 0 \cdot 20 = 1 \cdot 70 \text{ m}$$

Assuming cone base angle as 45°, hopper bottom angle as 50°, and free board, $H_5 =$ 0.15 m.

$$H_3 = 0 \cdot 5\,(1 \cdot 70 - 0 \cdot 15) \tan 50 = 0 \cdot 92 \text{ m}$$

and $$H_4 = 0 \cdot 5 \times 1 \cdot 30 \tan 45° = 0 \cdot 65 \text{ m}$$

Also $$H_1 = H_2 + H_4 + H_5 = H_2 + 0 \cdot 65 + 0 \cdot 15 = H_2 + 0 \cdot 80$$

The volume of holding the grain (V_g) can be determined by deducting the plenum chamber volume (V_2) from volume of the outer jacket and hopper bottom (V_1). Hopper top area is,

$$A_1 = \left(\frac{\pi}{4}\right) (1 \cdot 70)^2 = 2 \cdot 27 \text{ m}^2$$

and area of grain outlet is,

$$A_3 = \left(\frac{\pi}{4}\right) (0 \cdot 15)^2 = 0 \cdot 018 \text{ m}^2$$

then $V_1 = A_1 H_1 + \left(\frac{2H_3}{3}\right) [A_1 + A_3 + (A_1 A_3)^{0.5}]$

$$= 2\cdot27\,(H_2 + 0\cdot80) + \left(\frac{2\times0\cdot92}{3}\right)$$
$$\times[2\cdot27 + 0\cdot018 + (2\cdot27\times0\cdot018)^{0\cdot5}]$$
$$= 2\cdot27\,H_2 + 3\cdot35$$

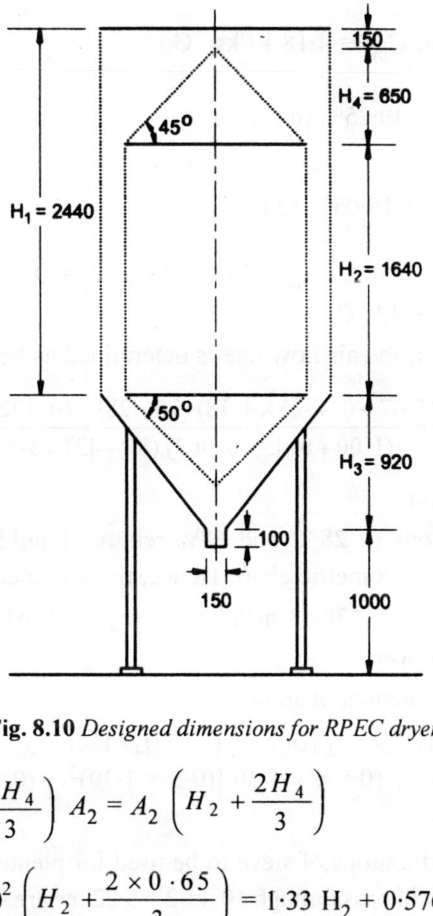

Fig. 8.10 *Designed dimensions for RPEC dryer*

and $V_2 = A_2 H_2 + \left(\frac{2H_4}{3}\right) A_2 = A_2 \left(H_2 + \frac{2H_4}{3}\right)$

$$= \left(\frac{\pi}{4}\right)(1\cdot30)^2 \left(H_2 + \frac{2\times0\cdot65}{3}\right) = 1\cdot33\ H_2 + 0\cdot576$$

then $V_g = V_1 - V_2$

$$= 2\cdot27\,H_2 + 3\cdot35 - (1\cdot35\,H_2 + 0\cdot576)$$
$$= 0\cdot94\,H_2 + 2\cdot77$$

Volume of 2.5 t paddy grain will be,

$$V_{gp} = \frac{2500}{580} = 4\cdot31\ \text{m}^3$$

but $V_g = V_{gp}$

or $0{\cdot}94\, H_2 + 2{\cdot}77 = 4{\cdot}31$

then $H_2 = 1{\cdot}64$ m

and $H_1 = 1{\cdot}64 + 0{\cdot}80 = 2{\cdot}44$ m

For determination of air flow rate requirement, the following parameters are considered.

$$W_d = \frac{2500\,(100-16)}{100} = 2100 \text{ kg}$$

$$C_{pp} = 1{\cdot}67 \text{ kJ/kg}, \; C_{pw} = 4{\cdot}18 \text{ kJ/kg °C}$$

$$X_1 = \frac{(33 \times 100)}{(100-33)} = 49{\cdot}25\% \text{ (d.b.)}$$

and
$$X_2 = \frac{(16 \times 100)}{(100-16)} = 19{\cdot}05\% \text{ (d.b.)}$$

$$T_{p1} = 28 \text{ °C}, \; T_{p2} = 70 \text{ °C}, \; \lambda_w = 2335 \text{ kJ/kg}, \; H_1 = 0{\cdot}067 \text{ kJ/kg}$$
$$T_{a1} = 87 \text{ °C}, \; T_{a2} = 42 \text{ °C}.$$

Using equation (8.20) the air flow rate is determined as below,

$$m_a = \frac{1{\cdot}1 \times 2100[(1{\cdot}67 + 0{\cdot}4925 \times 4{\cdot}18)\,(70-28) + (0{\cdot}4925 - 0{\cdot}1905) \times 2335]}{(1{\cdot}00 + 0{\cdot}45 \times 0{\cdot}067)\,(87-42) \times 3{\cdot}5 \times 60}$$

$$= 204{\cdot}5 \text{ kg/min}$$

At ambient conditions of 28°C and 68% relative humidity the humid volume is $0{\cdot}873 \text{ m}^3/\text{kg}$ from the psychrometric chart. Hence, the volumetric air flow rate will be,

$$V_a = 204{\cdot}5 \times 0{\cdot}873 = 178{\cdot}53 \text{ m}^3/\text{min} \qquad \text{say, } 180 \text{ m}^3/\text{min}$$

Power requirement for blower

The surface area of plenum chamber is,

$$= \pi D_2 H_2 + 2\,\{0{\cdot}5\, \pi D_2\, [(0{\cdot}5\, D_2)^2 + (H_4)^2]^{0{\cdot}5}\}$$
$$= 1{\cdot}30\pi \times 1{\cdot}64 + 2\,\{0{\cdot}5\, \pi \times 1.30\, [(0{\cdot}5 \times 1{\cdot}30)^2 + (0{\cdot}65)^2]^{0{\cdot}5}\}$$
$$= 10{\cdot}45 \text{ m}^2$$

Considering the perforations of sieve to be used for plenum chamber as 50% of the total surface area, the air will pass through $10{\cdot}45/2 = 5{\cdot}22 \text{ m}^2$ area. Then, the air requirement of the dryer per unit area of plenum chamber is, $183/5{\cdot}22 = 35{\cdot}06 \text{ m}^3/\text{min/m}^2$. For 0·305 m grain depth the pressure drop is 9·24 cm of water for $36{\cdot}50 \text{ m}^3/\text{min/m}^2$ air flow rate from Shedd's curve (Agril. Engineering Handbook). By interpolation for 0·20 m the pressure drop will be,

$$\Delta p = \frac{9 \cdot 24 \times 0 \cdot 20}{0 \cdot 305} = 6{\cdot}06 \text{ cm of water}$$

Considering the density of air as 1.13 kg/m³, the pressure drop in terms of air column is,

$$\Delta p = \left(\frac{6 \cdot 06}{100}\right)\left(\frac{1000}{1 \cdot 13}\right) = 53.63 \text{ m of water}$$

and the power requirement will be,

$$P = \frac{53 \cdot 63 \times 204 \cdot 5 \times 9 \cdot 81}{59176} = 1 \cdot 82 \text{ kW}$$

Hence, provide 2·24 kW (3·0 *hp*) electric motor.

Example 8.2 : *For the parameters given in Example 8.1, design the LSU type dryer with grain residence time of 30 min and air velocity in port as 4·80 m/s.*

Solution. Under the similar conditions of parameters the heat and mass balance in the LSU dryer will be the same. Thus, V_a = 18 m³/min.

Provide 1·50 m square drying chamber with height H_2 (Fig. 8.11). Then, the volume of drying chamber will be,

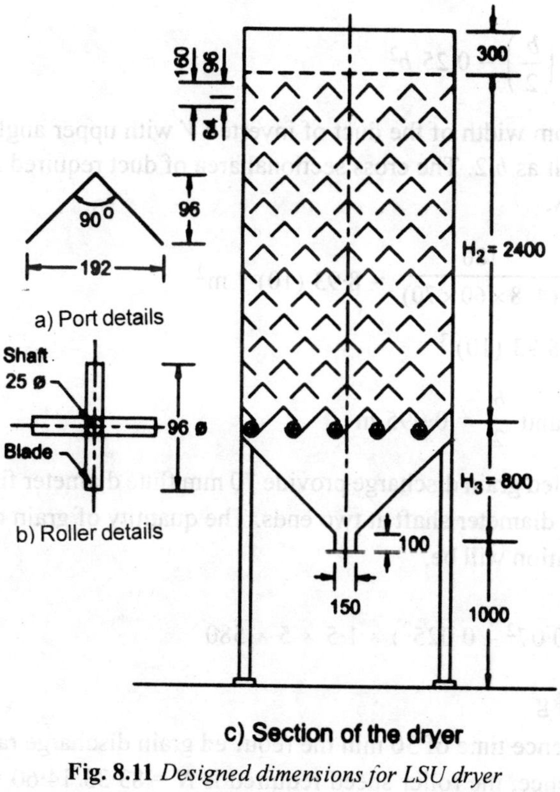

Fig. 8.11 *Designed dimensions for LSU dryer*

$$V_1 = (1{\cdot}5)^2 H_2 = 2{\cdot}25\, H_2$$

and the volume of ducts will be,

$$V_2 = \frac{V_a B}{v} = \frac{180 \times 1{\cdot}5}{(4{\cdot}8 \times 60)} = 0{\cdot}94 \text{ m}^3$$

then, the volume of drying chamber occupied by grain will be,

$$V_g = V_1 - V_2 = 2{\cdot}25\, H_2 - 0{\cdot}94 = 4{\cdot}31$$

or $\quad H_2 = 2{\cdot}33$ m

Provide the port pitch as 0·30 m and row-to-row pitch as 0·16 m, $H_2 = 2{\cdot}40$ m to accommodate the full number of rows *i.e.* 15. Let, grain outlet size be 0·15 m square then,

$$H_3 = 0{\cdot}5\, (1{\cdot}50 - 0{\cdot}15) \tan 50 = 0{\cdot}80 \text{ m}$$

and $\quad H_1 = H_2 + H_3 = 2{\cdot}40 + 0{\cdot}80 = 3{\cdot}20$ m

In addition to this, provide 0·30 m free board above the air ports. Width of the dryer will accommodate five (1·5/0·3) ports in each row. The discharge rollers will be accommodated in the lowest row and the total number of function ports for grain drying will be, $n = 14 \times 5 = 70$. The cross sectional area for each duct (A) will be,

$$A = (0{\cdot}5\, b)\left(\frac{b}{2}\right) = 0{\cdot}25\, b^2$$

where, b is the bottom width of the duct of inverted V with upper angle of 90° and lower 45° each with, height as $b/2$. The cross sectional area of duct required for the drying air is determined as below,

$$A = \frac{V_a}{(vn)} = \frac{180}{(4{\cdot}8 \times 60 \times 70)} = 8{\cdot}93\, (10)^{-3} \text{ m}^2$$

then $\quad 0{\cdot}25\, b^2 = 8{\cdot}93\, (10)^{-3}$

or $\quad b = 0{\cdot}19$ m and $\dfrac{b}{2} = 0{\cdot}095$ m

For the controlled grain discharge provide 70 mm flute diameter five discharge rollers mounted on 25 mm diameter shaft at two ends. The quantity of grain discharged by these rollers in one revolution will be,

$$G = \left(\frac{\pi}{4}\right) (0{\cdot}07^2 - 0{\cdot}025^2) \times 1{\cdot}5 \times 5 \times 580$$

$$= 14{\cdot}60 \text{ kg}$$

With the residence time of 30 min the required grain discharge rate is, $G_d = 2500/30 = 83{\cdot}33$ kg/min. Hence, the roller speed required is $N = 83{\cdot}33/14{\cdot}60 = 5{\cdot}7$ rpm.

***Example 8.3 :** A 10 tpd pigeonpea milling industry requires a LSU type dryer with 5 t holding capacity to be dried within three hours. The raw grain is stored at 32 °C and 60% relative humidity at 18% (wb) moisture content which is to be reduced to 10% (wb) for dehulling. Heated air shell enter the plenum chamber at 100 °C and shall leave the dryer at 48 °C. The grain has angle of repose of 32° and leaves the dryer at 80 °C. The grain residence time in the dryer and air velocity in the port shall be restricted to 25 min and 5 m/s, respectively.*

Solution. $W_d = \dfrac{5000\,(100-10)}{100} = 4500$ kg

$$X_1 = \frac{(18 \times 100)}{(100-18)} = 21 \cdot 95\% \text{ (d.b.)}$$

and $$X_2 = \frac{(10 \times 100)}{(100-10)} = 11 \cdot 11\% \text{ (d.b.)}$$

$T_{p1} = 32$ °C, $T_{p2} = 80$ °C, $T_{a1} = 100$ °C $T_{a2} = 48$ °C

Quantity of heat required for drying of pigeonpea grain is,

$$q_d = 1{\cdot}10 \times 4500\,[1{\cdot}67 + 0{\cdot}2195 \times 4{\cdot}18)\,(80-32) + (0{\cdot}2195 - 0{\cdot}1111) \times 2309{\cdot}5]$$
$$= 1854021{\cdot}5 \text{ kJ}$$

and heat to be given by the heated air

$$q_a = (1{\cdot}00 + 0{\cdot}45 \times 0{\cdot}075)\,(100-48) \times 3 \times 60\, m_a$$
$$= 9675{\cdot}9\, m_a$$

Hence, the rate of air flow is,

$$m_a = \frac{1854021{\cdot}5}{9675{\cdot}9} = 191{\cdot}61 \text{ kg/min}$$

From psychrometric chart the humid volume $V_h = 0{\cdot}889$ m^3/kg for storage condition of 32 °C and 60% relative humidity.

Then $V_a = 191{\cdot}61 \times 0{\cdot}889 = 170$ m^3/min

Assuming the 1·60 m square drying chamber with height H_2,

$$V_1 = (1{\cdot}6)^2\, H_2 = 2{\cdot}56\, H_2$$

and $$V_2 = \frac{170 \times 1{\cdot}60}{(5 \times 60)} = 0{\cdot}91 \text{ m}^3$$

then $V_g = V_1 - V_2 = 2{\cdot}56\, H_2 - 0{\cdot}91$

but volume of grain with 800 kg/m^3 bulk density is,

$$V_g = \frac{5000}{800} = 6{\cdot}25 \text{ m}^3$$

then $2{\cdot}56\ H_2 - 0{\cdot}91 = 6{\cdot}25$

or $H_2 = 2{\cdot}79$ m say, 2·80 m

The port pitch of 40 cm and row pitch of 20 cm will accommodate four ports in a row and 14 rows in the drying chamber; the total being 13 × 4 = 52 in operation since the last row is left for providing the discharge rollers. With hopper angle of 55° and outlet size 15 cm square,

$$H_3 = 0{\cdot}5\ (1{\cdot}60 - 0{\cdot}15) \tan 45° = 0{\cdot}72 \text{ m}$$

then $H_1 = 2{\cdot}80 + 0{\cdot}72 = 3{\cdot}52$ m

Cross sectional area of each duct,

$$A = 0{\cdot}5\ b \left(\frac{b}{2}\right) = 0{\cdot}25\ b^2$$

Also, this area is given by air flow requirement, that is,

$$A = \frac{170}{(5 \times 60 \times 60)} = 0{\cdot}011 \text{ m}^2$$

then $0{\cdot}25\ b^2 = 0{\cdot}01$

and $b = 0{\cdot}21$ m

Then, the height of duct will be 0·5 *b* *i.e.* 0·105 m.

Assuming four discharge rollers mounted on 25 mm diameter shaft, one for each duct with 80 mm flute diameter, the grain quantity discharged is,

$$G = \left(\frac{\pi}{4}\right) (0{\cdot}08^2 - 0{\cdot}025^2) \times 1{\cdot}60 \times 4 \times 800$$

$$= 23{\cdot}22 \text{ kg}$$

The discharge rate required for 25 min residence time will be,

$$G_d = \frac{5000}{25} = 200 \text{ kg/min}$$

and speed of the roller will be, $N = \dfrac{200}{23{\cdot}22} = 8{\cdot}61$ rpm

Nomenclature

A	: area of heat transfer, m^2
A_d	: cross section area of dryer through which air passes, m^2
A_w	: water surface area, m^2
C_{pp}	: specific heat of product, kJ/kg °C
C_{pv}	: specific heat of vapour, kJ/kg °C

C_{pw} : specific heat of water, kJ/kg °C
C_s : humid heat of gas
C_{s1} : humid heat of gas at inlet, kJ/kg
D : diffusivity of water vapour through grain, m^2h
dm_v/dt : rate of moisture removal, kg/h
dX : change in moisture content, %
dX/dt : change in moisture content per unit time, %/h
H_a : absolute humidity at ambient temperature, kg/kg
H_i : humidity of air entering the dryer, kg/kg
H_o : heat capacity of drying air, kJ/kg
H_s : humidity of saturated air leaving the dryer, kg/kg
H_w : absolute humidity at temperature t_w, kg/kg
h : film heat transfer coefficient at air and water interface, kJ/m^2 ^0C
K_d : drying constant
K_w : mass transfer coefficient at air and water interface, kg/h m^2
m_a : mass flow rate of drying air, kg/min
m_a' : mass flow rate of drying air per unit area, kg/h m^2
m_g : mass flow rate of drying gas, kg/min
m_p : mass flow rate of product, kg/h
m_s : average mass transfer rate, kg/h
N_v : drying rate for unit surface, kg/h
N_w : drying rate for initial condition, kg/h
P : power requirement for blower, kW
p_a : partial pressure of air, m of water
p_w : partial pressure of water, m of water
Δp : pressure drop of air flow, m of water
Q : change in enthalpy of drying material, kJ/h
q : heat loss due to radiation and other reasons, kJ
q_a : heat capacity of drying air, kJ/min
q_d : heat required for drying of product, kJ
q_1 : sensible heat of product, kJ
q_2 : sensible heat of water, kJ
q_3 : latent heat of water evaporation, kJ
R : radius of spherical particles, mm
R_c : drying of grain under constant rate period, kg/h
R_{ch} : characteristic drying rate, kg/h

R_f	: drying of grain under falling rate period, kg/h
T_a	: drying air temperature, °C
T_{a1}	: drying air temperature at inlet, °C
T_{a2}	: drying air temperature at exhaust, °C
T_{g1}	: drying gas temperature at inlet, °C
T_{g2}	: drying gas temperature at exhaust, °C
T_{p1}	: initial temperature of product, °C
T_{p2}	: final temperature of product, °C
T_v	: vaporization temperature, °C
T_{v1}	: initial vaporization temperature, °C
T_{v2}	: final temperature of vapour, °C
t	: drying time at any instant, min
t_c	: drying time for constant rate period, h
t_{d1}	: time of advance for drying front, h
t_{d2}	: drying time after reaching the drying front at top, h
t_f	: drying time for falling rate period, h
V_a	: volumetric flow of air, m^3/min
v_h	: humid volume of air, m^3/kg
W_d	: bone dry weight of grain to be dried, kg
X	: grain moisture content at drying time t, %
X_a	: average grain moisture content at the end of drying front reaching at the top layer, %
X_c	: critical grain moisture content at characteristic drying rate, %
X_e	: equilibrium grain moisture content, %
X_1	: initial grain moisture content, %
X_2	: final grain moisture content, %
x_1	: lower limit of grain moisture content, %
x_2	: higher limit of grain moisture content, %
ϕ_{ch}	: moisture ratio at characteristic drying rate
λ_w	: latent heat of water evaporation, kJ/kg

9

Spray Dryers

9.1 Introduction

The heat sensitive liquids can be dried in a single operation in spray dryers by atomizing it using a nozzle or rotating disc inside the closed chamber and stream of hot air. Spray drying has a rapid drying cycle, short retention time (8 to 10 s) of product in drying chamber giving out directly the final product. Use of high drying air temperatures (150 to 600 °C) is possible due to the fact that the product droplets are never be at higher temperature than the wet bulb temperature of drying air. Thus, short exposure of the product and low product temperature do not cause damage to the heat sensitive products such as milk and fruit juices. Spray drying effectively controls the product properties and qualities with large quantity production continuously using simple equipment at high efficiency. It produces relatively uniform and spherical product particles. Since, the product contact with inner surface of drying chamber is in anhydrous condition, the corrosion problem is not simplified. However, spray drying is not suitable for high density product and is not flexible in use. A unit designed for fine atomization may not be able to produce a course product and vice versa. It has large evaporation rates and the raw product must be pumpable. Product recovery and dried product collection increase the initial and operating costs.

9.2 Construction

A typical spray dryer illustrated in Fig. 9.1 consists of a cylindrical drying chamber with conical bottom and is properly insulated around to avoid the heat loss. The raw liquid filtered through the strainer is fed by pumping to the atomizer situated at the top centre of the drying chamber which atomizes the liquid feed into fine droplets. The drying air heated to a required high temperature enters the drying chamber either at the top or bottom, mixed with the fine droplets and moisture is evaporated immediately within a short period. Dried particles are collected at the bottom and are taken out either by a rotary valve or screw conveyor. The hot air passes through a cyclone separator where any entrained dried particles are separated. There may be two or three cyclone separators in series depending on the particle size. Sometimes a bag filter is also provided if the particles are very fine. The exhaust fan imparts the draft to the hot air entering and leaving the drying chamber through the cyclone separators.

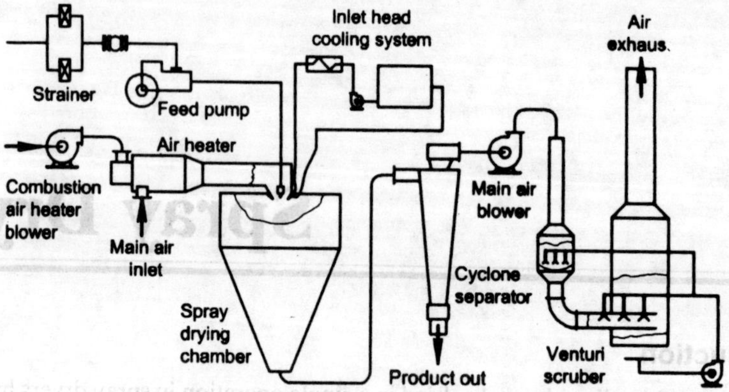

Fig. 9.1 *Typical spray dryer and system*

9.3 Classification

Spray dryers have been classified in three groups based on the material and air flow patterns, namely, counter-current flow, co-current flow and mixed flow dryers (Fig. 9.2). The counter-current flow dryer is shown in Fig. 9.3 in which the liquid feed in atomized at the top centre of the drying chamber and droplets fall down. The hot air is introduced near the bottom of the chamber and while moving up mixes with the droplet flow to achieve the drying. Air is removed from the top of the chamber where as the product leaves the chamber and while moving up mixes with the droplet flow to achieve the drying. Air is removed from the top of the chamber where as the product leaves the chamber from bottom. Figure 9.4 illustrates a co-current spray dryer in which hot intake air and fine droplets are mixed at the atomizer. Both move down when moisture evaporation takes place. The product and air leave the drying chamber from bottom exit. The co-current flow may occur in three patterns in addition to that shown in Fig. 9.4 *i.e.*, downward parallel flow, mixed type flow and upward flow. In case of parallel flow pattern (Fig. 9.5) the product and air flow in uniform lines from top to the bottom of a narrow drying chamber. Both together leave the chamber for further separation. This allows to use a high velocity (2 to 3 m/s) and temperature as compared with those for the co-current type pattern. Figure 9.6 illustrates a mixed flow pattern in which the fluid is atomized at some location near the centre of the drying chamber. Air is introduced at the top which moves down and contacts the fine droplets. Then it moves upward towards the air outlet provided at the top of the chamber. The product leaves the exit port near the bottom of the chamber. Although, this system has high evaporating capacity per unit volume, high air temperature may affect the product quality. In case of upward flow pattern the hot air is introduced at the bottom of drying chamber near the point of atomization.

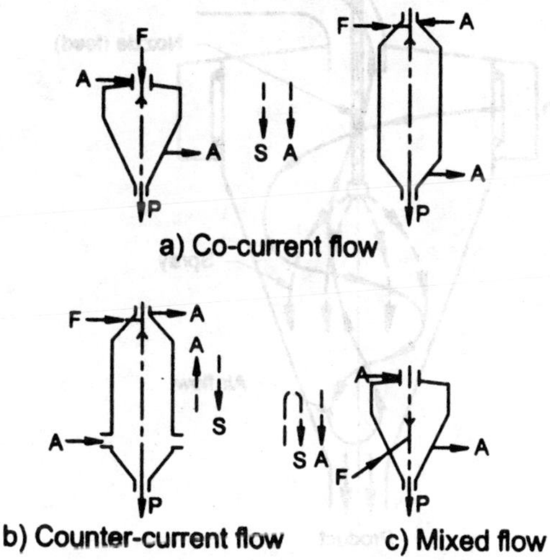

Fig. 9.2 *Types of spray dryers*

[A : air, F : feed, P : product, S = spray]

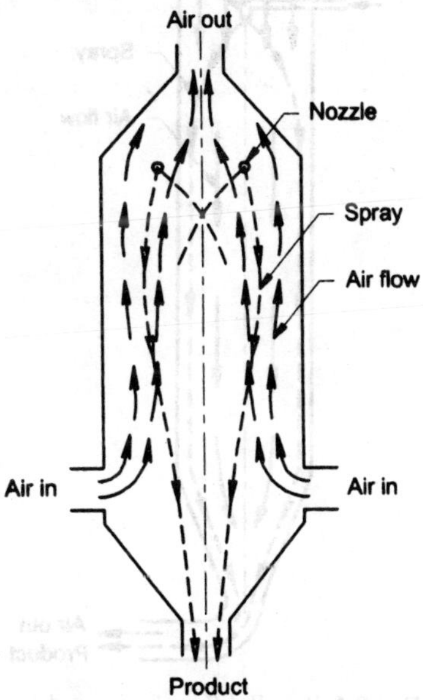

Fig. 9.3 *Counter-current flow in spray dryer*

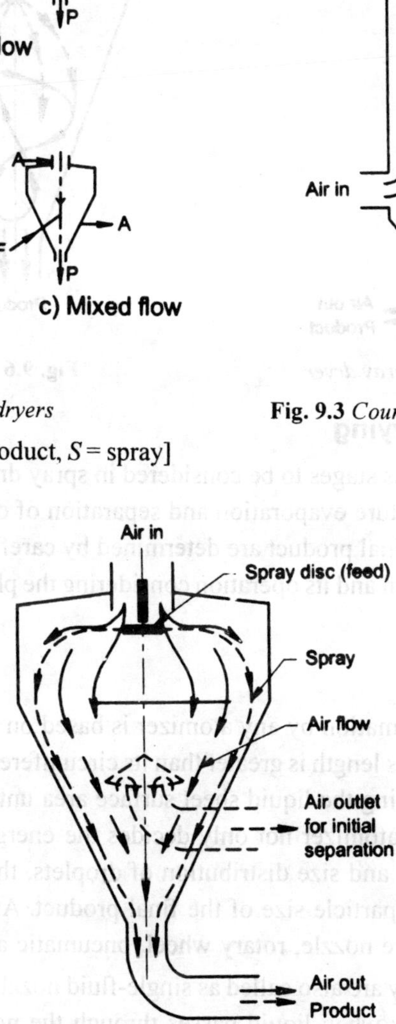

Fig. 9.4 *Cocurrent flow in spray dryer*

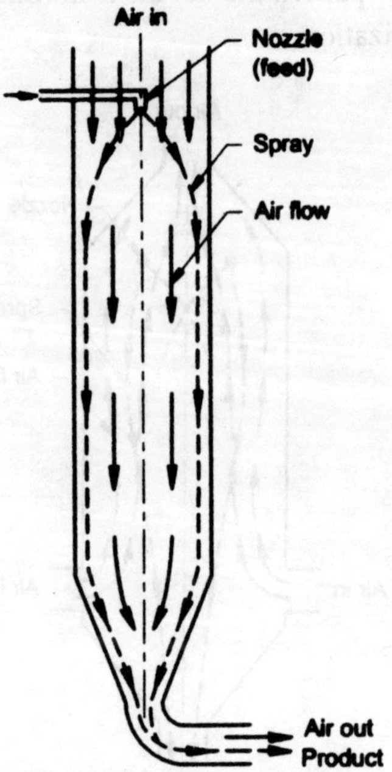

Fig. 9.5 *Parallel flow in spray dryer*

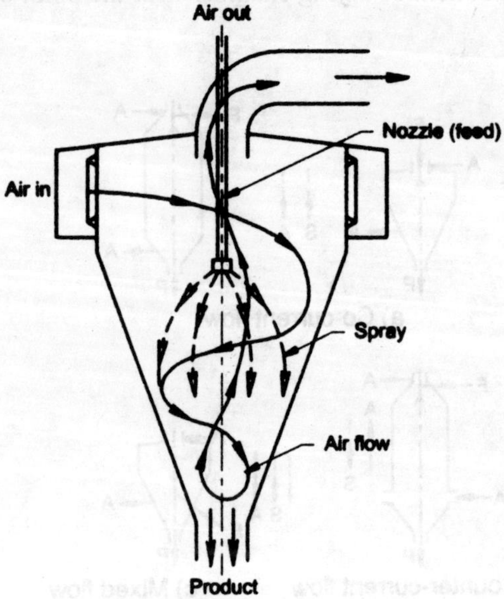

Fig. 9.6 *Mixed flow in spray dryer*

9.4 Stages in spray drying

There are three process stages to be considered in spray drying, namely, atomization, spray air mixing with moisture evaporation and separation of dried product from exit air. The characteristics of the final product are determined by carefully conducting each stage according to the dryer design and its operation considering the physico-chemical properties of the liquid feed.

9.4.1 Atomization

The liquid droplet formation by any atomizer is based on the theory that a free cone of liquid is unstable when its length is greater than its circumference. The droplet formation can be achieved by increasing the liquid sheet surface area until it becomes unstable and disintegrates. The type of atomizer not only decides the energy requirement for droplet formation but also the size and size distribution of droplets, their velocity and trajectory which, in turn, affects the particle size of the final product. Atomizers or nozzles are of four types, namely, pressure nozzle, rotary wheel, pneumatic and airless spray nozzle.

Pressure nozzles: They are also called as single-fluid nozzle (Fig. 9.7) which converts pressure to velocity energy when liquid passes through the nozzle at about 5 to 7 MPa pressure. Entry of liquid is tangential and exit is in the form of a hollow cone, the angle being 40 to 140°. Their maximum capacity is rated to 100 lph and the orifice diameter range is 0·4 to 4·0 mm. Number of nozzles can be provided for higher capacity. Narrow and tall drying chambers are required for these nozzles. The dried particles are usually hollow spheres with narrow range of the size. Pressure nozzles are not recommended for highly concentrated suspensions and abrasive materials, since, they can clog or erode the nozzle orifice. They consume less energy as compared with the other types of nozzles.

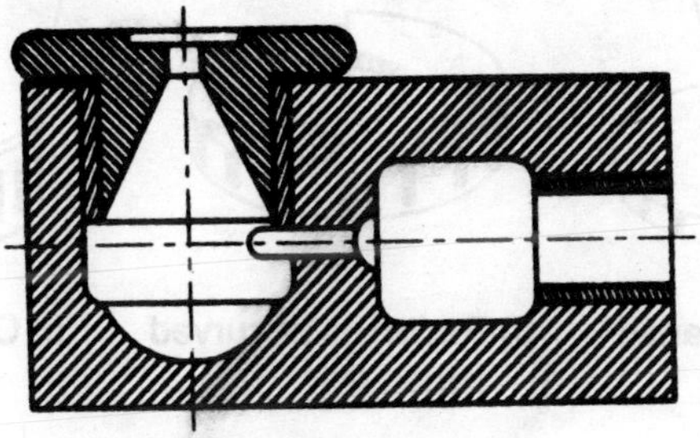

Fig. 9.7 *Pressure nozzle*

The average droplet diameter can be computed roughly from pressure drop through the nozzle using the following impiricol relationship.

$$d = 9575\ (\Delta p)^{0{\cdot}33}$$

Rotary wheel nozzles: These nozzles are provided with as shown in Fig. 9.8 (*a*), (*b*) and (*c*) and a rotary disc mounted on shaft. Liquid is fed at the centre of a thin sheet at the periphery of the disc and ultimately disintegrated into the feed rates up to 200 lph and properties. Usually, 10000 to 30000 rpm (100 to 200 m/s) disc speed is necessary to create desired droplet size. The product quality and capacity of the spray dryer is decided by the number and shape of the vanes. The number or height of vanes is increased to achieve higher capacity. Figure 9.8 (*d*) shows the simple flat disc rotating at high speed, widely used for drying the fruit juices. In milk industry, curved vanes are provided [Fig. 9.8 (*e*)] to produce a high density powder. The rotary wheels are generally driven by the electric motors either directly or by belt drive through helical or epicyclic gear systems.

For quick and rough estimates of droplet size, the following empirical expression should be used once it does not contain any physical parameter on liquid side,

$$d = 0{\cdot}241 \left[\frac{1}{N_s}\right]^{0{\cdot}6} \left[\frac{1}{\rho_f}\right]^{0{\cdot}3} \frac{\mu m_f}{2 d_s \rho_f} \cdot \frac{\sigma}{N_v b} \qquad (9.3)$$

Pneumatic nozzles: They use compressed air or steam for atomization of liquid feed and hence, known as two-fluid nozzles (Fig. 9.9). In most cases, the liquid feed is mixed with air outside the nozzle body. About 0·5 m^3 compressed air is needed per kilogram of fluid. The maximum capacity per nozzle is one tph of fluid feed. The spray angle ranges between 20 and 60° depending on the nozzle design. The droplet size is controlled by air-fluid ratio and gives high degree of homogeneity to the product. These atomizers use the highest amount of energy for operation.

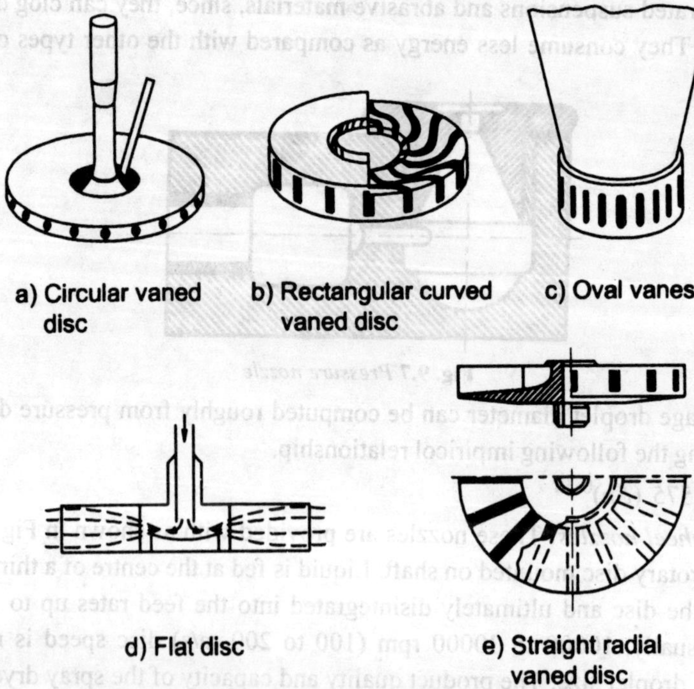

Fig. 9.8 *Rotary wheel atomizer*

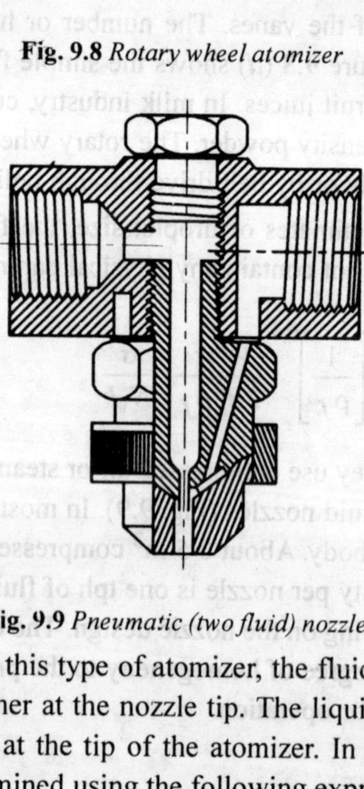

Fig. 9.9 *Pneumatic (two fluid) nozzle*

Airless spray nozzles: In this type of atomizer, the fluid is divided into two streams and again brought back together at the nozzle tip. The liquid stream breaks up into fine droplets due to impingement at the tip of the atomizer. In case of these atomizers, the droplet diameter can be determined using the following expression.

$$d = \frac{5\cdot 85(10)^5}{v_r}\left[\frac{\sigma}{\rho_f}\right]^{0.5} + 597\left[\frac{\mu}{(\sigma\rho_f)^{0.5}}\right]^{0.45}\left[\frac{1000V_f}{V_a}\right]^{1.5} \quad (9.4)$$

The high pressure of liquid is suddenly reduced at atmospheric pressure at the tip of atomizer and when the liquid splashes in the form of discrete droplets, the specific surface area increases almost 1000 times. The droplets travel after formation, till they are completely dry by evaporation of moisture. Evaporation time depends on the liquid being sprayed and the relative saturation of air inside the drying chamber. The shape and size of the drying chamber depends upon the type of atomizer used and its angle of spray. Temperature profile in the drying chamber should be carefully studied particularly, for heat-sensitive products and the diameter and height of the chamber should be such that the largest droplet in the spray is completely dried before it reaches the chamber wall in order to avoid built up of dried product on wall surface.

9.4.2 Spray-air mixing and dryer calculation

Drying of droplets inside the drying chamber takes place due to simultaneous heat and mass transfer between the droplet and the surrounding air. Initially, droplet surface water is dried at constant rate and as the solid concentration increases, droplet diameter is established. The droplet temperature rises and evaporation takes place from inside of the particle at falling rate. Initial small heating period is neglected. Initial feed material temperature is taken equal to the wet bulb temperature at which the moisture evaporation takes place. Heat is required for three-fold function in the spray dryer namely, heating the wet feed material, moisture evaporation from the droplet and drying the droplet to particle up to desired temperature. The drying time of the droplet can be determined by using the following expression.

$$t_d = \frac{q}{h A \theta_{lm}} \quad (9.5)$$

Surface heat conductance can be computed for droplets using the Nusselt number as below,

$$h = \frac{N_{Nu} k}{d_p} \quad (9.6)$$

Number of particles passing through the dryer per second is given by,

$$n = \frac{m_f}{(\pi/6) d_p^3 \rho_p \times 3600} \quad (9.7)$$

and the total surface area of these particles is,

$$A = \pi n d_p^2 \tag{9.8}$$

The velocity of deposition of particles in the spray dryer can be calculated by the use of a graph between Lyash Chenko number (N_{Ly}) and Archimedes number (N_{Ar}) with following relationship.

$$v_t = \left[\frac{N_{Ly} \nu \rho_f g}{\rho_a}\right]^{0.33} \tag{9.9}$$

The value of N_{Ly} is found from graph with N_{Ar} which is given by,

$$N_{Ar} = \frac{d_p^2 \rho_f g}{\nu^2 \rho_a} \tag{9.10}$$

The particle velocity obtained from equation (9.9) is the theoretical velocity and the actual value is taken as 10 to 25% higher than theoretical value. Considering the drying time and particle velocity, the cylindrical height of the drying chamber can be determined as below.

$$H = t_d (v_p - v_t) \tag{9.11}$$

When the particles enter the dryer tube, they are not able to accelerate immediately. Therefore, the height of the dryer determined from equation (9.11) may be increased by 1·6 to 2·0 times. Diameter of the drying chamber can be estimated as,

$$D = \left[\frac{4Va}{(\pi v_p)}\right]^{0.5} \tag{9.12}$$

The volumetric flow rate of drying air can be assessed from the heat balance in the dryer.

9.4.3 Discharge of product and air

The dried product is usually collected at the bottom of the drying chamber and discharged. The product air separation may either takes place inside or outside the drying chamber. The fine particles can be separated in a cyclone separator or separators in series. Figure 9.10 illustrates the discharging systems. Product is separated out of the chamber from bottom [Fig. 9.10 (*a*)] while it is accumulated inside the chamber [Fig. 9.10 (*b*)] through [Fig. 9.10(*d*)] and discharged by means of valves or rotating scrapers. Sometimes air sweepers may be provided to discharge the dried product along the drying chamber wall.

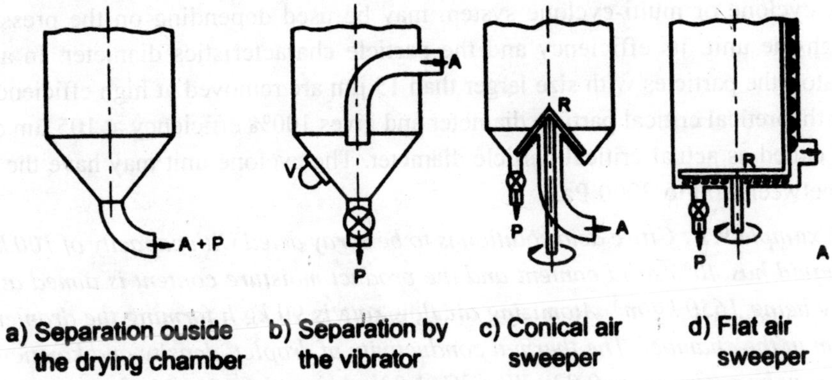

Fig. 9.10 *Product discharging systems*

[A : air outlet, P : dry product discharge, R : air sweeper, V: vibrator]

9.5 Ancillary devices

The spray drying process decides the ancillary devices to be provided with the system. They are drying air heaters, air blowers and powder separators.

Drying air heaters: Generally, direct air heater are widely used since, they have a simple design, more supply of natural gas and can provide air temperature upto 800 °C. The consumption rate gas or fuel used for heating can be expressed as,

$$m_g = \eta_h \left[\frac{m_a C_{pa}(T_o - T_i)}{q_c} \right] \tag{9.13}$$

Air blowers: Usually, centrifugal blowers are employed in the spray drying process where high air flow are required. There may be either a single blower system (the whole system is under high negative pressure) or a double blower system (one blower provided after the cyclone separator and second at the inlet duct of drying chamber). This arrangement facilitates ease in pressure control inside the chamber. Impeller blade design dictates the pressure developed in the chamber/system which may be back-curved, radial or intermediate between these two. Back-curved blades may be deposit the particles on blade backs when powder air ratio is high. The other two have better self cleaning characteristics. Energy consumption by the blower can be determined by the following equation,

$$E = \frac{\Delta_p V_a}{\eta_b} \tag{9.14}$$

Powder separators: The maximum quantity of dried powder must be separated by the cyclone separator which employs a centrifugal force to move the product particles towards the outer periphery of the separation chamber and to separate them from air swirling around. The particles are collected from the outlet at the bottom and air leaves at top. A single cyclone or multi-cyclone system may be used depending on the pressure drop through the unit, its efficiency and the particle characteristics diameter. In a cyclone separator, the particles with size larger than 15 μm are removed at high efficiency, called as the theoretical critical particle diameter and gives 100% efficiency at 105 μm diameter, being called as actual critical particle diameter. The cyclone unit may have the pressure drop between 700 to 2000 Pa.

Example 9.1 : *Citirc acid solution is to be spray dried with capacity of 100 kg/h. The feed liquid has 40% solid content and the product moisture content is aimed at 3%, the density being 1650 kg/m³. Atomizing air flow rate is 90 kg/h forming the droplets of size 100 μm in the chamber. The thermal conductivity of droplet, density and kinetic viscosity of air may be assumed as 0·028 W/m °C, 1·03kg/m³ and 6·681 $(10)^{-5}$ m²/s, respectively. Latent heat of vaporization of moisture is 2520 kJ/kg. The specific heat and temperature of feed material, product and air are given below.*

	Product	*Specific heat, kJ/kg °C*	*Temperature, °C*
1.	*Feed*	*3·85*	*65*
2.	*Product*	*1·64*	*60*
3.	*Drying air*	*1·02*	*160*
4.	*Ambient air*	*1·01*	*30*
5.	*Exhaust air*	—	*70*
6.	*Water*	*4·18*	

If the wet bulb temperature inside the drying chamber is 45°C, determine the drying time and chamber dimensions.

Solution. Considering a mass balance of the system,

$m_f X_f = m_p X_p$ and $m_f = m_p + m_w$ for feed, product and moisture,

$$m_w = m_f - \frac{m_f X_f}{X_p} = 100 - \frac{100 \times 0 \cdot 40}{1 - 0 \cdot 03} = 58 \cdot 76 \text{ kg/h}$$

Similarly, energy balance of the system is,

(*A*) Heat input requirement in the system is three fold

(*a*) Heating the wet feed material

$$q_f = m_f C_{pf} (T_w - T_a)$$
$$= 100 \times 3 \cdot 85 \,(45 - 30)$$
$$= 5775 \text{ kJ/h}$$

(*b*) Evaporation of moisture

$$q_w = m_w \lambda_w = 58{\cdot}76 \times 2395 = 140730 \text{ kJ/h}$$

(*c*) Drying of product to 60°C

$$q_p = m_p C_{pp} (T_p - T_w)$$
$$= (100 - 58{\cdot}76) \times 1{\cdot}64 (60 - 45)$$
$$= 1015 \text{ kJ/h}$$

(*d*) Heat to be supplied for atomization of air

$$q_a = m_a C_{pa} (T_p - T_a)$$
$$= 90 \times 1{\cdot}01 (60 - 30)$$
$$= 2727 \text{ kJ/h}$$

Then, the total heat requirement will be,

$$q_{in} = q_f + q_w + q_p + q_a$$
$$= 5775 + 140730 + 1015 + 2727 = 150247 \text{ kJ/h}$$

(*B*) This heat will be equal to the enthalpy of the product and exhaust air

(*a*) Enthalpy of the product

$$q_p = m_p C_{pp} (T_p - T_w) = (100 \times 0{\cdot}40) \times 1{\cdot}64 (60 - 45) = 984 \text{ kJ}$$

(*b*) Enthalpy of exhaust air

$$q_d = m_d C_{pd} (T_d - T_e)$$
$$= m_d \times 1{\cdot}02 (60 - 70)$$
$$= 91{\cdot}8\, m_d$$

Then, the heat balance of the complete system yields,

$$91{\cdot}8\, m_d = 150247$$

or $\quad m_d = 1636{\cdot}7$ kg/h or $27{\cdot}3$ kg/min

Specific volume of air at 160°C is 1·23 m^3/kg and its density at 30°C is 1·13 kg/m^3, hence, the volumetric flow rate of drying air will be,

$$V_d = 27{\cdot}3 \times 1{\cdot}23 = 33{\cdot}58 \text{ m}^3\text{/min or } 0{\cdot}56 \text{ m}^3\text{/s}$$

Considering the Nusselt number as 2·0 for very small droplet size of 100 μm from equation (9.6), surface heat conductance is,

$$h = \frac{2 \times 0{\cdot}028}{[100(10)^{-6}]} = 560 \text{ W/m}^2\text{ °C}$$

Number of particles passing through the dryer per second is,

$$n = \frac{m_f}{(\pi/6) \times d^3 \times \rho_f \times 3600}$$

$$= \frac{m_f}{(\pi/6)\times[100(10)^{-6}]^3 \times 1650 \times 3600}$$

$$= 32152513{\cdot}76$$

Total surface area of droplets is,

$$A = n\,\pi d^2 = 32152513{\cdot}76\ \pi\ [100\ (10)^{-6}]^2 = 1{\cdot}01\ \text{m}^2$$

also, log mean temperature difference from equation (5.1) is,

$$\theta_{lm} = \frac{(160-30)-(70-60)}{\ln[(160-30)/(70-60)]} = 46{\cdot}78$$

Then, from equation (9.5) the time of drying will be,

$$t_d = \frac{150247 \times 1000}{560 \times 1{\cdot}01 \times 46{\cdot}78 \times 3600} = 1{\cdot}58\text{s}$$

From equation (9.10), Archimedes number is given as,

$$N_{Ar} = \frac{[100(10)^{-6}]^2 \times 1650 \times 9{\cdot}81}{[6{\cdot}681(10)^{-5}]^2 \times 1{\cdot}03}$$

For this N_{Ar} = 35207, from graph N_{Ly} = 205, then the theoretical velocity of particle from equation (9.9) will be,

$$v_t = \left[\frac{205 \times 6{\cdot}81(10)^{-5} \times 1650 \times 9{\cdot}81}{1{\cdot}03}\right]^{0{\cdot}33} = 5{\cdot}88\ \text{m/s}$$

Let, particle velocity be 15% higher than v_t then

$$v_p = 5{\cdot}88 \times 1{\cdot}15 = 6{\cdot}76\ \text{m/s}$$

Considering the drying time and particle velocity, the cylindrical height of the drying chamber will be,

$$H_t = 1{\cdot}86\ (6{\cdot}76 - 5{\cdot}88) = 1{\cdot}64\ \text{m}$$

Actual height can be taken 80% higher than H_t, then,

$$H = 1{\cdot}64 \times 1{\cdot}80 = 2{\cdot}95\ \text{m say, } 3{\cdot}0\ \text{m}$$

and the diameter of the drying chamber from equation (9.12) will be,

$$D = \left[\frac{4V_d}{\pi v_p}\right]^{0{\cdot}5} = \left[\frac{4 \times 0{\cdot}34}{6{\cdot}76\pi}\right]^{0{\cdot}5} = 0{\cdot}25\ \text{m}$$

Nomenclature

A : total surface area of droplets passing the dryer per unit time, m^2/s
b : height of the vane, mm
C_{pa} : specific heat of air, kJ/kg °C
D : diameter of the drying chamber, mm
d : droplet diameter, m
d_p : particle diameter, m
d_s : diameter of the disc, mm
E : Energy consumption by the blower, W
g : gravitational constant, m/s^2
H : height of cylindrical drying chamber, m
h : surface heat conductance of droplet, kJ/hm^2
k : thermal conductivity of material, W/m °C
m_a : mass flow rate of air, kg/h
m_f : liquid (fluid) feed rate, kg/s
m_g : fuel consumption rate, kg/h
N_{Ar} : Archimedes number
N_{Ly} : Lyash Chenko number
N_{Nu} : Renolds number
N_s : speed of the disc, rps
N_v : number of vanes on the disc
n : number of particles passing through the dryer per second
Δp : pressure drop through the nozzle, Pa
q : total heat requirement for drying system, kJ
q_c : calorific value of gas or fuel, kJ/kg
T_i : air temperature at inlet, °C
T_o : air temperature at outlet, °C
t_d : drying time for the droplet, s
V_a : volumetric air flow rate, m^3/s
V_f : volumetric fluid flow rate, m^3/s
v_p : actual velocity of the particle, m/s
v_r : relative velocity of the particle, m/s
v_t : theoretical velocity of the particle, m/s

θ_{lm} : log mean temperature difference, °C

η_b : efficiency of the blower, %

η_h : efficiency of the heater, %

σ : fluid surface tension, N/m

μ : fluid viscosity, Ns/m^2

ν : kinetic viscosity of the air, Ns/m^2

ρ_a : density of air, kg/m^3

ρ_f : density of fluid feed, kg/m^3

ρ_p : density of particles, kg/m^3

10

Milling Equipments

10.1 Introduction

Hard bodies such as grains are subjected to milling in order to obtain free flowing material like flour which consists of uniform size particles. This material may either be a final product or to be further processed. The grinding process in this case is called plain or simple grinding. In some cases, non-homogeneous hard bodies are subjected to selective grinding in several stages, called complex grinding. At each stage the process is carried out in such a way that particles differing from one another according to one or the other physical property are obtained. This facilitates to separate a number of fractions from ground mixture having more or less homogenous particles. Wheat milling into graded flour in based on the difference between the mechano-structural properties of hull and endosperm. The brittle endosperm is crushed into parts while hull becomes flake-like so that their separation from endosperm parts is facilitated. With the selective method and optimum conditions for working tools of grinders and their kinematic parameters leads to the poor quality product and hence, also of the bread.

Heat is produced during grinding resulting into the low quality product. The output of the grinder is decreased with excessive grinding with higher specific power consumption and increased production costs.

10.2 Effectiveness of grinding

The degree or effectiveness of grinding is defined by the ratio of the overall surface of the particles of ground product to the overall surface of the particles of the initial product that is,

$$Ge = \frac{S_a}{S_b} \tag{10.1}$$

In flour milling, usually, i = 20 to 50 depending on the flour grade required. The surface of particles can be determined by screen analysis in a sifter comprising of number of screens with holes of various sizes. The product retained by each sieve is designated with the characteristics of screen e.g. the product passed through screen number 90 and retained on screen number 120 is designated as 90/120. The larger the number of fractions separated, the more complete and accurate the surface of the particles is determined. The number of particles in each fraction is given by,

$$n_i = \frac{W_i}{d^3 \rho} \tag{10.2}$$

The surface of the individual particle is assumed to be approximately equal to the surface of d cube of an equivalent size *i.e.* $6d^2$. where d, is the side of the cube equal to mean diameter of the product. Then the surface of a group of particles (fraction) is given by,

$$S_1 = n_1 \, 6 \, d^2 = \frac{6W_i}{d^3 \rho} \tag{10.3}$$

and overall surface of particles of whole sample before grinding is given by,

$$S_b = S_1 + S_2 + \ldots + S_m = 6 \sum_1^m \left(\frac{W}{d\rho} \right) \tag{10.4}$$

10.3 Generalized law of grinding

Energy input for grinding is used for formation of elastic and plastic deformations and for overcoming the forces of molecular attraction. Subsequently, the body gets disintegrated resulting into particles having larger overall surface. The body absorbs energy when subjected to the elastic deformation. On interruption of external forces, the body returns only part of energy absorbed, and part converting into thermal energy, which results in the rise of product temperature. During grinding the part of energy is consumed to charge the working surfaces of tools electrostatically and frictional resistance of machine elements.

According to Rebinder, the expenditure of energy on grinding any material into a product consisting of the particles of a given disparity is expressed in the following form,

$$E_t = E_d + E_s \tag{10.5}$$

Starting from this generalized law of grinding, the efficiency of a given process can be expressed as below,

$$\eta = \frac{E_s}{(E_d + E_s)} \tag{10.6}$$

This generalized law of grinding can be given in the detailed form as,

$$E_t = E_p + N_d\left(\frac{\sigma_b^2 V}{2E}\right) + E_f(S_a - S_b)G_e^n \tag{10.7}$$

From equation (10.6), the efficiency is given by the expression,

$$\eta = \frac{E_f(S_a - S_b)G_e^n}{E_p + N_d(\sigma_b^2 V/2E) + E_f(S_a - S_b)G_e^n} \tag{10.8}$$

10.4 Roller mill

Roller mill consists of two cylindrical rollers as working tools revolving at differential speed in opposite direction. Grain or parts of grain are ruptured in a wedgelike space, which gradually narrows toward the bottom. This space is formed by the two cylindrical surfaces with an external contact surface. The grain starts to rupture some what above the line connecting the centres of the rolls. The slow roll which is usually the lower, holds the grain during the action of the surface of fast roller. The grain or its parts which have fallen into the grinding zone are subjected to simultaneous deformation by compression due to wedge shaped clearance and by shear due to differential roller speeds.

Grinding process of wheat consists of four stages, namely, formation, to middlings, their purification, reduction of the purified middlings, and separation of bran. In the first stage, the degree of grinding is light, extricating the maximum amount of coarse particles of endosperm, called middlings and contain small amount of bran. In the second stage, pure middlings are extracted from the mixture of ground particles and the fine particles are separated without any considerable grinding from the remaining mixture. The endosperm is reduced in the third stage to flour of the appropriate degree of fineness. In the fourth and final stage, coarse flake-like bran particles are obtained which have a minimum content of endosperm.

Other conditions being equal, the degree of grinding depends mainly on the ratio between the dimensions of the product cross section being ground and the clearance between rollers. It is also affected by the structural and mechanical properties of grain e.g., rupturing stress, optimum moisture content is grain, plastic and elastic deformation of grain: geometrical parameters of roller *i.e.*, diameter, macro and micro-geometry of the rollers, kinematic parameters of rollers and specific load on the rollers.

Roller clearance

Roller clearance is a very sensitive and critical parameter such that a very slight variation in it leads to a considerable variation in the extraction. The following relationship between extraction and roller clearance gives a good approximation to the results,

$$E_x = a \cdot e^{-cd} \tag{10.9}$$

Roller geometry

Other conditions being equal, the roller diameter, clearance and initial size of particle to be ground determine the conditions of rupture of the product. The relationship between the three parameters (Fig. 10.1) determine the complete path of the particle during grinding, which is approximately equal to length *BC*. The length of a section of path where primary rupture of particles into finer ones occurs depends on these three parameters. Considering a triangle O_2 *BC*, the length of path is given by,

$$L = [(0{\cdot}5\ D + 0{\cdot}5\ d)^2 - (0{\cdot}5\ D + 0{\cdot}5\ b)^2]^{0{\cdot}5}$$

or
$$L = [0{\cdot}5\ D\ (d - b) + 0{\cdot}25\ (d^2 - b^2)]^{0{\cdot}5} \tag{10.10}$$

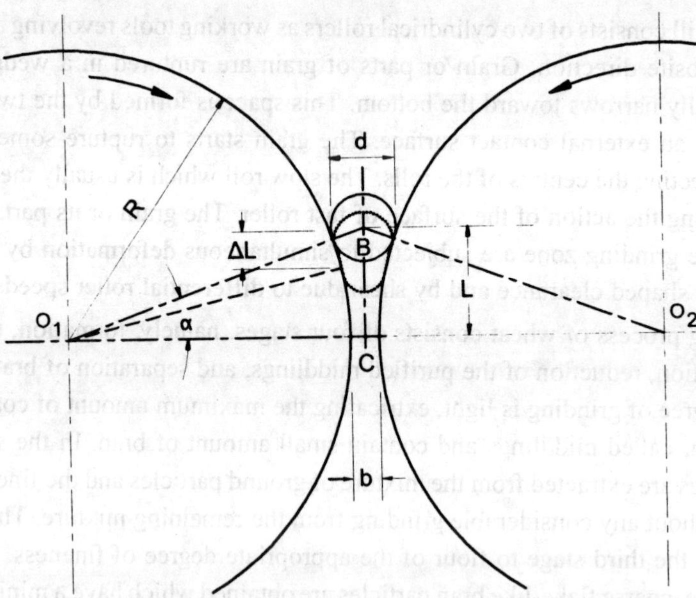

Fig. 10.1 *Roller geometry and grinding zone*

It is obvious from equation (10.10) that when the roller clearence and particle size remain constant, the path of the particle increases with the increase in roller diameter. The increase in degree of grinding was experimentally demonstrated by Lomonosov and explained that it was due to the fact that when the roller diameter is reduced the rate of deformation is increased since, there is a decrease in the sectional length ΔL along with the primary rupture of particles into inner ones occurs.

The corrugations on roller have two unequal lateral edges: *abcd*, a narrow one called sharp edge of *abgf*, a wide one called blunt edge, both forming the sharpening angle (Fig. 10.2). This angle θ is divided into two by a perpendicular line to the axis of roller through point a, the sharp angle α and blunt angle β. The obtuse angle ϕ included between the tangent *AB* and the line passing through point a *i.e.*, the apex of the corrugation and the front edge *abcd* is termed as cutting angle. The distance t_1 along the periphery between two consecutive apexes is called as the corrugation pitch. The radial distance or height of corrugation h is the distance between apex and hollow edge of the corrugations. The pitch and number of corrugations are connected by the following equation.

$$p_t = \frac{\pi D}{N_c} \quad (10.11)$$

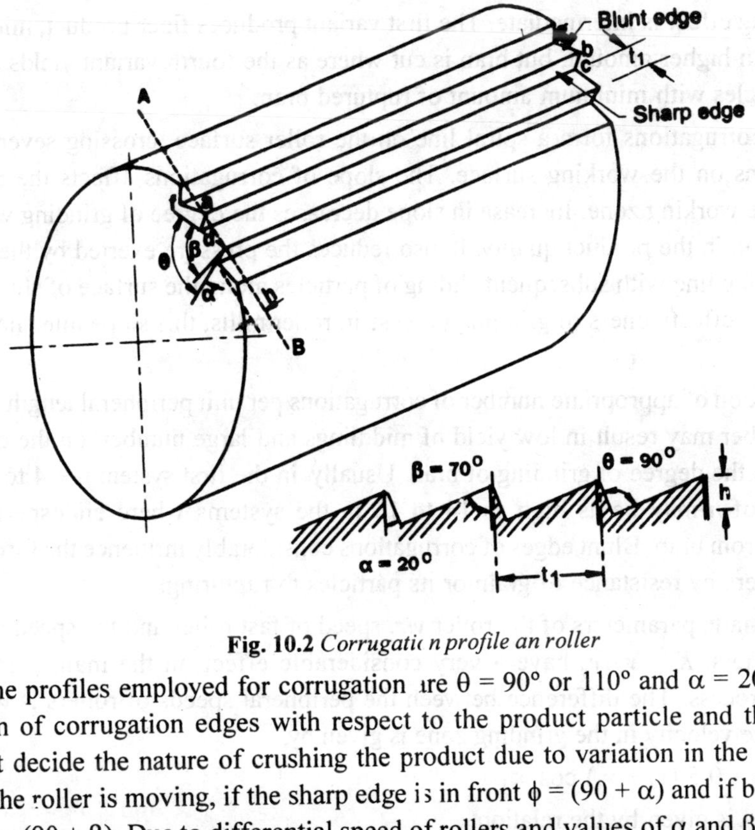

Fig. 10.2 *Corrugatic n profile an roller*

The profiles employed for corrugation are θ = 90° or 110° and α = 20° or 30°. The position of corrugation edges with respect to the product particle and their action or product decide the nature of crushing the product due to variation in the cutting angle. When the roller is moving, if the sharp edge is in front ϕ = (90 + α) and if blunt edge is in from θ = (90 + β). Due to differential speed of rollers and values of α and β, four variants are possible for the corrugation positions with respect to each other.

In the first variant, sharp edge of the fast roller and blunt edge of slow roller are at front (sharp and sharp). As soon as the grain enters the grinding zone, the edges of both rollers cut into the grain while blunt edges of slow roller hold the grain back. In second variant, both blunt edges are at front (blunt to sharp). The sharp edges of the slow roller cut into grain which is held back by the pressure exerted due to blunt edges of the fast roller. With the advancement of other blunt edges of the fast roller, the slow roller glides along the grain shearing the half of the caryopsis adjacent to it. In the third variant, the sharp edges are in front of both the rollers (sharp to blunt). The sharp edges of the corrugations of the fast roller cut into the grain and carry it to the side of the line connecting the centres of the rollers. On the other side, the grain glides along the blunt and that of the slow roller is sharp (blunt to blunt). In such a position the grain its paretically compressed at the beginning. As the other blunt edge of the fast roller advances with respect to slow roller, the grain is shifted along the blunt edges of slow roller.

Among the four variants, most drastic grinding occurs in the "sharp to sharp" position, since, both rollers are active on the product. The mildest action is achieved in "blunt to blunt" position where no roller is in action for product deformation. In other two positions the grinding effect is intermediate. The first variant produces finer product, middlings are produced in higher amount, but bran is cut where as the fourth variant yields coarse and blunt particles with minimum amount of ruptured bran.

The corrugations form a spiral line on the roller surface, crossing several pairs of corrugations on the working surface. The slope of corrugations affects the position of grain in the working zone. Increase in slope decreases the degree of grinding without any deterioration in the product quality. It also reduces the pressure exerted by the fast roller on the cutting line with subsequent sliding of particles along the surface of slow roller. To increase the effectiveness of grinding process in roller mills, this slope must not exceed 4 to 6%.

Selection of appropriate number of corrugations per unit peripheral length of roller as small number may result in low yield of middlings and large number, on the other hand, increase in the degree of grinding of bran. Usually, in the first system n = 4 to 5 whereas the value of n may be as high as 14 to 15 in the systems where endosperm is to be separated from bran. Blunt edges of corrugations considerably influence the forces exerted on the rollers by resistance of grain or its particles to rupturing.

Kinematic parameters of the roller *viz.* speed of fast roller and the speed ratio of the pair of rollers, $K = v_f / v_s$ have a very considerable effect on the main parameters of grinding process. The difference between the peripheral speeds of rollers is $v_d = v_f - v_s$ and average velocity in the grinding zone is given by,

$$v_a = 0{\cdot}5\,(v_f + v_s)\cos \propto_1 \tag{10.12}$$

where $\propto_1$ is is given by the relation,

$$\cos \propto_1 = \frac{(D+b)}{(D+d)} \tag{10.13}$$

It should be noted that mutual position of corrugations affects the value of v_a *i.e.*, somewhat higher in case of "sharp to sharp" position and lower when they are in "blunt to blunt" position than predicated by equation (10.12). The holding time of product in grinding zone is given by $t = L$, the number indicating the action of the corrugations is given by the formula as below.

$$N_c = (v_f - v_a)\, tn_c \tag{10.14}$$

or $$N_c = (v_f - v_a)\frac{Ln_c}{v_a} \tag{10.14}$$

Considering the angle of contact being very small, cos $\propto_1$ may be unity. Substituting from K, equation (10.14) becomes,

$$N_c = \left[\frac{v_f}{v_s} - 1\right]Ln_c = \left[\frac{2Kv_s}{Kv_s + v_s} - 1\right]Ln_c = \left[\frac{2K - K - 1}{K+1}\right]Ln_c$$

or $$N_c = \left[\frac{K-1}{K+1}\right]Ln_c \tag{10.15}$$

The capacity of roller mill is the amount of product in units of weight which is ground in unit time given by,

$$q = 3{\cdot}6\ \rho\ L_r\ v_a\ b \tag{10.16}$$

Considering a coefficient of filling of the grinding zone, the theoretical capacity is given as,

$$q_t = \frac{q}{\psi} \tag{10.17}$$

The value of ψ in flour mills is 0·15 to 0·80, the lower related to first breaking system and higher to the last ones, mainly due to wider roller clearance (1·8 to 0·4 mm) in initial stages, higher density of the product with smaller number of particles per unit volume.

Nomenclature

a, c	: imperial coefficients
b	: roller clearance, mm
D	: diameter of the roller, mm
d	: mean diameter of product particle to be ground, mm
E	: modulus of elasticity, N/mm^2
E_d	: energy required for elastic and plastic deformation, Wh
E_f	: energy required per unit area of particles newly formed, Wh
E_p	: energy required for deformation processes and wear of grinding surfaces, Wh
E_s	: energy required for formation of new surfaces, Wh
E_t	: total energy required for grinding material into product particles, Wh
E_x	: extraction
G_e	: degree or effectiveness of grinding
K	: speed ratio for pair of rollers
L	: length of particle path during grinding, m

L_r : length of particle roller, m

ΔL : length of section of particle path where primary particle rupture into finer ones occurs during grinding, m

m : number of fractions from screens

N_c : number of corrugations on roller surface

N_d : number of deformation cycles for particles

n : number depending on grinding conditions

n_c : number of corrugations per unit peripheral length of the roller

n_i : number of particles in each fraction

p_c : corrugation pitch, mm

q : capacity of roller mill, kg/h

q_t : theoretical capacity of roller mill, kg/h

R : radius of the roller, mm

S_a : total surface of the particles after grinding, mm^2

S_b : total surface of the particles before grinding, mm^2

S_i : surface of a group of particles in fraction, mm^2

$S_{1+..+}S_n$: sum of surface of particles in each fraction, mm^2

t : holding time for product in grinding zone, s

V : volume of the material, mm^3

v_a : average velocity of particle in grinding zone, m/s

v_d : difference between the peripheral speed of rollers, m/s

v_f : peripheral speed of the fast roller, m/s

v_s : peripheral speed of the slow roller, m/s

W_i : yield weight of each fraction, g

α : sharp angle of grinding teeth, deg

α_1 : angle of contact between particles to be ground and roller, deg

β : blunt angle of grinding teeth, deg

θ : angle of sharpening of corrugation between sharp and blunt edges, deg

η : efficiency of grinding operation, %

ϕ : cutting angle, deg

ρ : bulk density of fraction, kg/m^3

σ_b : breaking stress of material to be ground, N/mm^3

ψ : coefficient of filling of the grinding zone

11

Belt Conveyers

11.1 Introduction

Belt conveyer has become a most popular means of transporting large quantities of materials over a very large distance at a low cost. It finds its extensive application on agricultural farms, fertilizer factories, large grain storages, processing plants, power plants, civil engineering fields, coal mines and also in the assembly bay of small scale manufacturing industries. It is able to transport the material in varied forms such as powder, granular or lumpy etc. between operations in production flow lines. Even with a highly abrasive materials like sand, coke or crushed stone etc., a belt conveyer with the proper type of belt with proper loading and unloading facilities, careful maintenance and servicing, out lasts 2 to 3 times the chain or bucket elevator because fine abrasives of the dust work into the chain joints grinding down the pins and brushes. It has been noticed that if the production is balanced and a vast quantity is out turned, then perhaps no other single equipment is so economical for handling material as a single belt conveyer.

11.2 Constructional features

Essentially, a belt conveyer consists of five components as shown in Fig. 11.1, namely, (*a*) an endless belt to carry the material and to transmit the pull, (*b*) belt supports for material carrying side, called as upper idlers which may be flat or troughed and on return run, called lower idlers, (*c*) driving unit comprising of a prime mover, driving drum or drums, driving gears or chains etc., (*d*) supporting structure for all the components and (*e*) accessories to maintain belt tension such as gravity take-up, loading and discharging devices like hoppers, chutes, trippers etc. and belt cleaners.

A continuous endless canvas or rubber belt is arranged on two pulleys at the ends of a belt conveyer : the take-up pulley at the loading side and driving pulley at the discharge head. The belt is supported at intermediate points by rollers or idlers at closer spacing on load carrying side than those on return run. The former may be straight or troughed but

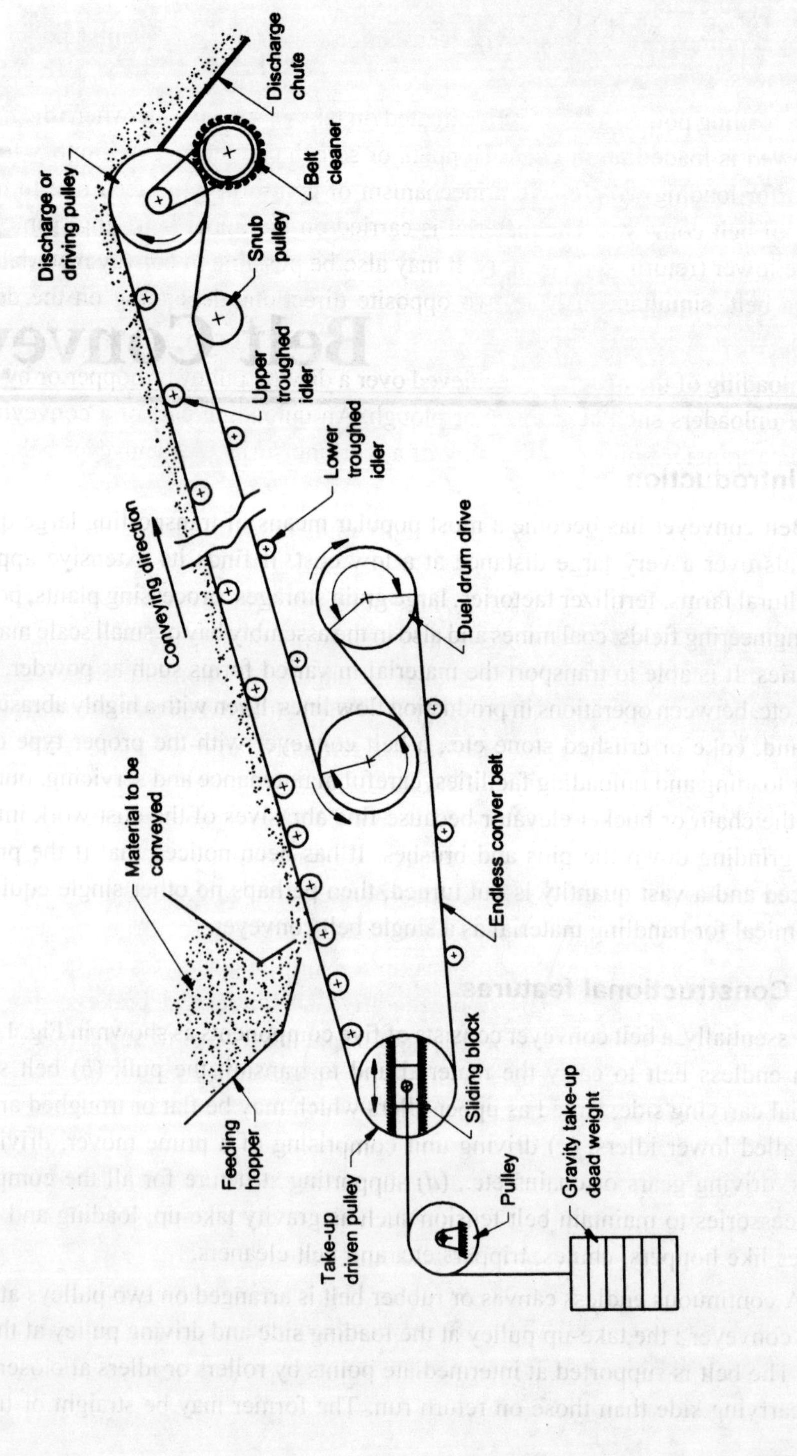

Fig. 11.1 *Schematic details for belt conveyer*

later are usually straight. The idlers are mounted on the supporting frame and the belt bends around the take-up and driving pulleys and rolls over the idlers. The driven pulley is rotated by a prime mover, usually an electric motor in processing plants and an engine at remote places.

The loading points are generally situated on take-up pulley side where the material to be conveyed is loaded on the belt. Hoppers or screen conveyers or vibratory feeders are provided for loading which have a mechanism of controlling the feed rate to match the capacity of belt conveyer. The material is carried on the upper (carrying) run of the belt while the lower (return) run is empty. It may also be possible to convey materials on both runs of a belt, simultaneously in two opposite directions depending on the demand of situation.

Unloading of the material is achieved over a driving pulley or hopper or by means of a special unloaders such as trippers or plough. An unloading part of a conveying belt is termed as a transfer point. For cleaning of a working surface of conveyer belt, a cleaner is provided at the starting point of the return run.

Belt conveyers can carry materials on horizontal or inclined path, the maximum angle of inclination being limited from 15 to 23° depending upon the dynamic coefficient of friction of load against the belt surface, angle of repose of bulk material, method of loading and belt speed. The bulk material can retain well (without slippage) if the angle of incline of conveyer belt is about 10 to 15° less than the angle of friction of material at rest on the belt. This margin is necessary due to sagging of the belt between two successive idlers so that incline angle at roller support may be substantially larger than the required average geometrical angle of conveyer incline. Moreover, belt vibrations promote sliding-down of the material which is elevated at higher belt speeds. A greater angle of inclination is possible for continuous uniform loading as compared with interrupted periodic loading. In case of inclined conveyers, it is necessary to incorporate some safety devices like electrically operated brakes or ratchet plate type hold-back to prevent accidents in case of power failure or at the end of the work. Besides, in a retarding conveyers (down-hills material conveying), an electrical dynamic, regenerative or centrifugal brake system must be provided, as otherwise, belt speed that of the prime mover due to acceleration of gravity causing the damage either to the later or the transmission system.

11.3 Classification

Belt conveyers can be classified according to their path profile, mobility, application and method of transfer of tractive force. According to the path profile belt conveyers can be classified as horizontal [Fig. 11.2 (*a*)] and inclined [Fig. 11.2 (*b*)]. They can also be mixed types such as inclined-horizontal [Fig. 11.2 (*c*)], horizontal-inclined [Fig. 11.2 (*d*) and with intricate path [Fig. 11.2 (*e*). Conveyers should be designed for simple and straight path. All conveyers can be divided into the stationary and mobile and further into portable, extensible or floating (on water). According to their design and application, conveyers can be classified into general-purpose and special-ones, like underground, food conveyers in food and feed industries, conveyers employed in flow lines in the instrument making, radio engineering and light industries. Considering the method of tractive force transmission, the belt conveyers may be tractive (used for both load-carrying and tractive element) and cable-belt and chain-belt conveyers (where the belt functions only as the load-carrying element and two cables or chains serve as the tractive element.

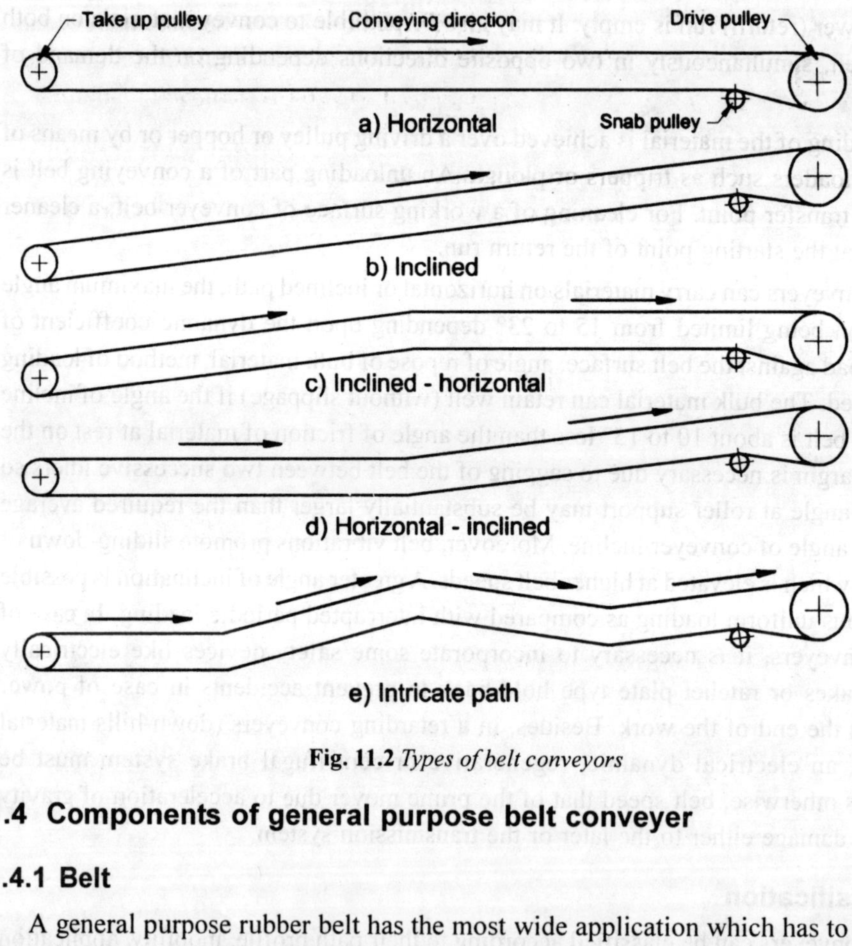

Fig. 11.2 *Types of belt conveyors*

11.4 Components of general purpose belt conveyer

11.4.1 Belt

A general purpose rubber belt has the most wide application which has to carry two types of load namely, the weight of the material to be conveyed and the tractive load. In order to withstand these loads, a belt should have high strength, elasticity in both the directions, resistance to alternating loads on pulleys and rollers and to abrasion. It should also have low mass and low relative elongation under the load. A conveyer belt is made up of a tractive carcass covered completely by an elastic filler made of synthetic rubber. The carcass makes the belt stronger in both the directions to take longitudinal and transverse stresses (Fig. 11.3) and the filler protects carcass from mechanical damage, abrasion and atmospheric conditions. The belts may be rubber fabric or rubber-cable distinguished on the basis of carcass. Fabric plies, usually of carbon, nylon, lavsan and other synthetic materials or even cotton threads of high strength are provided in rubber-fabric type. The threads are arranged in a single or double row along the belt length and crosswise too, running parallel to one another. The number of plies may be 2 to 14 depending on its width, strength required and the transverse rigidity. The later is the ability of a conveyer belt to form a trough on roller supports, for which the number of plies for standard width is specified. For a given tension in belt, the number of plies can be determined by the following relationship.

$$N = \frac{C_f \cdot P_{max}}{\sigma_n B} \tag{11.1}$$

The value of C_f is dependent on the number of plies and the belt width. The ranges are as below.

N	2-4	4-5	6-8	9-11	11-14			
C_f	9.0	9.5	10.0	10.5	11.0			
B, mm	400	500	600	800	1000	1200	1400	1500
C_f	2-4	3-5	3-6	4-8	5-9	5-10	6-11	8-12

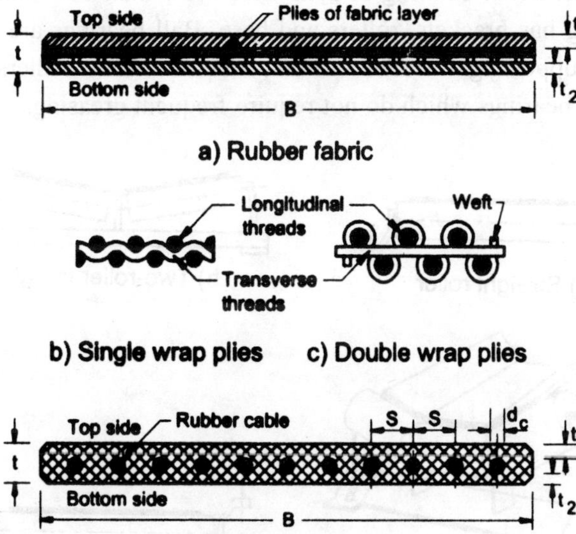

a) Rubber fabric

b) Single wrap plies c) Double wrap plies

d) Rubber cable belt

Fig. 11.3 *Construction of conveyer belt*

For common fabric belt σ_n = 45-60 N/mm per ply, for extra strong cotton duck fabric or synthetic fabric σ_n = 80-120 N/mm per ply. Rubber-cable belts have the carcass in the form of steel cables laid along the belt length and parallel to each other. Such belts may have σ_n= 800 N/mm per ply.

Rubber-fabric belt has the advantages of wide range for selection for variety of belt and its strength, joining of ends is possible by various methods (mechanical, vulcanization etc.), high resistance to the longitudinal breakage and a high shock-absorbing capacity. Its limitation is a large elongation (up to 4%) and require large size pulleys. Rubber-cable belt is characterized by a high strength, low elongation (up to 0·25%) and longer service life. Its drawbacks are a large mass and complicated belt joining at ends.

11.4.2 Roller supports

Belt running on take-up and drive pulleys is supported by rollers of various types and designs. According to their position on conveyer, roller supports are classified as upper idlers (supporting the load carrying side of belt) and lower idlers (supporting the return run). Upper idlers may be straight [Fig. 11.4 (*a*)] or troughed on two, three [Fig. 11.4 (*b*) and (*c*)] or even five rollers. Straight idlers are rarely used in the conveyer sections under plough unloaders, process flow-line conveyers or for loading or unloading of bags and packages. Troughed rollers are essentially used for conveying bulk loads such as grain, powder etc. Rubber discs, single or continuous, or rubber coatings are provided on upper idlers in loading section of belt in order to absorb the shocks. Lower idlers are mostly straight, single roller cylindrical [Fig. 11.4 (*d*)] or disc type or sometimes two-roller troughed at 10° inclication for better centring of belts wide more than 1000 mm. A roller support with rigid fastening has brackets, rollers and base. Ball bearings are usually fitted in a cylindrical sleeve supporting the roller axle on both ends. Modern idlers are provided with grease-packed ball bearings which do not require frequent greasing.

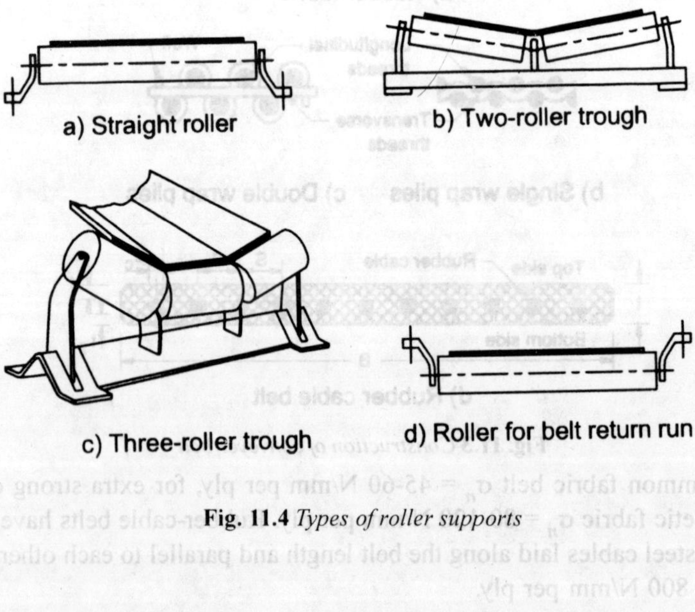

Fig. 11.4 *Types of roller supports*

The belt may move aside from the roller supports. When in operation due to unsymmetrical loading on it, sticking of material to idlers etc. For this reason, belt centralizing arrangement is provided in roller supports; which automatically return the belt to the central position. The rollers used for straight belt are provided with 1 to 3° inclination to the inward direction (Fig. 11.5) so that the belt is pushed toward the centre from both sides and belt remains in central position. In case of troughed belt an arrangement is made such that the three-roller support mounted on the frame can rotate along the vertical axis within certain limit (Fig. 11.6). When the belt moves aside, it passes against the side roller, turning the frame through certain angle with longitudinal direction. An opposite force is exerted on belt due to this slanting of roller support which tends to move the belt to the central position. After the return of belt to central position, the roller support is also turned-back into a normal position.

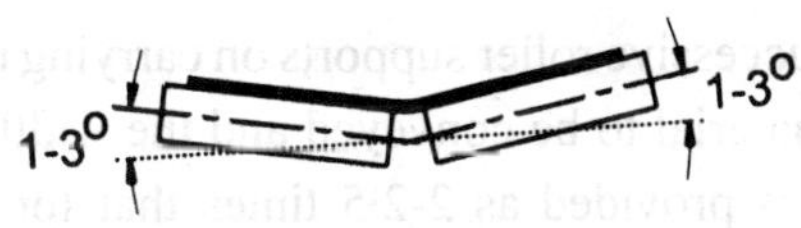

Fig. 11.5 *Arrangement for belt contralizing*

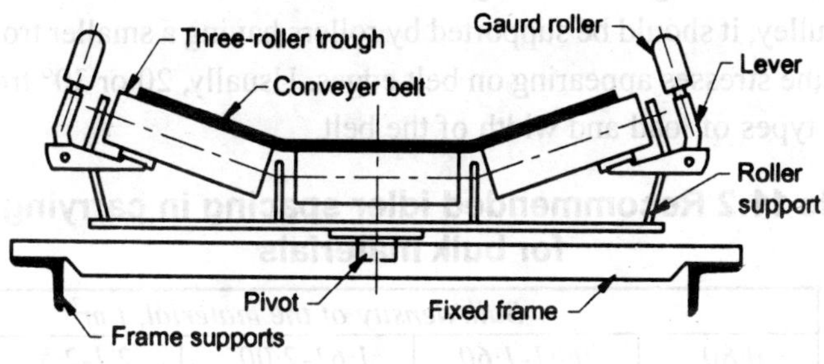

Fig. 11.6 *Roller support revoling assembly for belt centering*

The diameter of idler is dependent on the type of conveyer, density of the material to be conveyed and the width and speed of the belt (Table 11.1). The idler diameter increases with an increase in density and lump size, belt width and speed, higher coefficient of friction in bearings and with a lower coefficient of adhesion between belt and idler surfaces. The middle horizontal idler should be stronger than the side ones in a three-roller troughing support because about 70% of the total load is exerted on middle rollers and 15% on each of the side rollers.

Table 11.1 Recommended roller diameters for troughing and straight supports

Roller diameter, mm	*Belt width, mm*	*Maximum values of the Bulk density, t/m³*	*Maximum values of the Belt speed, m/s*
89	400, 500, 650, 800	1·6	2·0
108	400, 500, 650,	2·0	2·5
	800, 1000, 1200	1·6	2·5
133	800, 1000, 12000	2·0	2·5
159	800, 1000, 1200	3·5	4·0
	1400, 1600, 2000-3000	3·5	3·2
194, 219, 245	800, 1000, 1400	3·5	4·0
	1600, 2000, 3000	4·0	6·3

The spacing between successive roller supports on carrying run is provided depending on the bulk density of the material to be conveyed and the width of the belt (Table 11.2). On return run, the spacing is provided as 2-2·5 times that for the carrying run but not more than 3·5 m. In conveyers for lump or package carrying, at least 3-5 shock absorbing rollers are arranged in loading section; may be 0·4-0·5 m spaced. Where the belt changes its cross-section from troughed to straight and vice versa i.e. before the head pulley and after the drive pulley, it should be supported by rollers having a smaller troughing angle, in order to reduce the stresses appearing on belt edges. Usually, 20 or 30° troughing angle is provided for all types of load and width of the belt.

Table 11.2 Recommended idler spacing in carrying run for bulk materials

Belt width, mm	*Bulk density of the material, t/m³* <0·80	0·81-1·60	1·61-2·00	2·1-2·5	> 2·5
400	1·5	1·4	1·4	1·3	1·2
500	1·5	1·4	1·4	1·3	1·2
650	1·4	1·3	1·3	1·2	1·2
800	1·4	1·3	1·3	1·2	1·1
1000	1·3	1·2	1·2	1·1	1·1
1200	1·3	1·2	1·2	1·1	1·1
1400	1·3	1·2	1·2	1·1	1·1
1600	1·3	1·2	1·2	1·1	1·1
2000	1·3	1·2	1·1	1·0	1·0

11.4.3 The drives

The driving force is transmitted from pulley to the belt due to friction between the two surfaces in contact. The drive consists of a pulley, transmission mechanism and a prime mover. Drives of an inclined belt conveyer include a braking device which prevents slipping back of the loaded belt under the weight of the material conveyed, if the prime mover fails. Various types of drives are shown in Fig. 11.7. It may be of a single pulley, with or without snub pulley and twin pulley. Single pulley drives [Fig. 11.7 (*a*)] are used for low power (30-50 kW) transmission located at delivery end. An angle of wrap of 180° is passible in this drive. It may be increased up to 210° by providing a snub pulley. Double-pulley drive [Fig. 11.7 (*b*)] is used for heavy installations, usually at some convenient intermediate point between the head and drive pulleys. It has an angle of wrap of about 350° for both the pulleys. In case of twin-pulley tandem drive, the belt is wrapped around the two drums (pulleys) to increase a wrapping angle to the tune of about 400° for heavy drives.

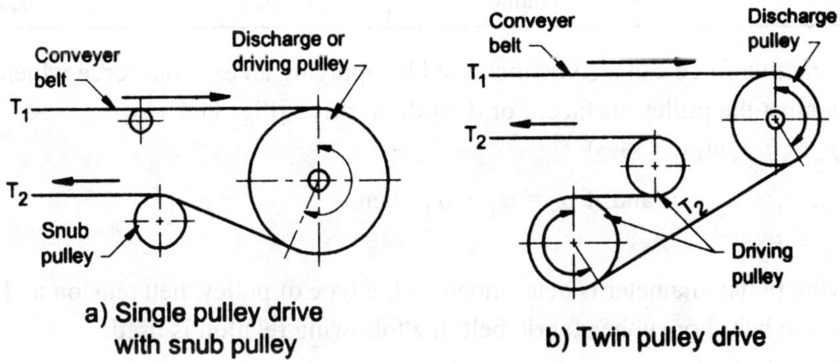

Fig. 11.7 *Types of belt conveyer drives*

The advantages of single pulley drive are simple in design, high reliability, compact, belt load reversal at only one point and return run of inclined and horizontal inclined conveyers can be conveniently used due to its gravity force for descending loads. The limitations are restricted wrapping angle (not more than 240°) and has low coefficient of utilization of the belt strength. Double-pulley drive has the merit over single-pulley drive of having greater total wrapping angle (up to 480°) due to which it is possible to use a less stronger belt. Their drawbacks are larger size, lower reliability of operation, higher cost, multiple bending of the belt and hence, shorter service life of the belt.

In order to utilize the full strength of the belt, the profile and length of the path and local conditions of operation, the coefficient of friction between the belt and pulley surfaces should be considered (Table 11.3). According to the general theory of single pulley drive the tensions in a belt section running on pulley *i.e.* on tight side and in that running off the pulley are interrelated as below.

$$P_1 \leq P_2 \, e^{\mu\alpha} \tag{11.2}$$

The term $e^{\mu\alpha}$, known as pull factor, determines the tractive capacity of a driving pulley. The circumferential tractive force that can be transmitted to pulley (neglecting losses due to belt rigidity) can be determined by the relationship as below.

$$P = P_1 - P_2 = P_2\,(e^{\mu\alpha} - 1) \qquad (11.3)$$

Thus, in order to improve tractive force to be transmitted, the friction coefficient can be raised by lining the pulley with rubber or providing grooves of 3-4 mm deep for dust collection and moisture depletion.

Table 11.3 Rated values of coefficient of friction for belt and pulley surfaces

Condition of the steel/cast iron pulley		*Coefficient of friction for pulley*	
Contacting surface	*Atmosphere*	*With lining*	*Without lining*
Clean	Dry	0.50	0.35
Dusty	Dry	0.40	0.30
Laden with granular material	Humid	0.25	0.20

The tractive force can also be increased by applying an external force which presses the belt against the pulley surface. For dual-drum drive [Fig. 11.7 (*b*)],

$$P_2 = P_3\, e^{(\mu_1\alpha_1 + \mu_2\alpha_2)} \qquad (11.4)$$

Practically, $\mu_1 = \mu_2 = \mu$ and if $\alpha_t = \alpha_1 + \alpha_2$, then,

$$P_2 = P_3 e^{\mu\alpha t} \qquad (11.5)$$

Driving pulley diameter is determined by the type of pulley, belt tension and number of plies in the belt. For rubber-fabric belt, the following relation is used.

$$D_p = C_d N \qquad (11.6)$$

where, C_d = 100 to 125 for N = 2 to 6 and C_d = 150 for N = 8 to 12. The return pulley diameter is taken as 80 to 100 times N and guide pulley diameter is provided as 50 to 70 times N. The selected diameter of drive pulley should be checked for the belt pressure acting on pulley surface which should not exceed the allowable stress.

$$\sigma_b = \frac{360\,(P_{s1} + P_{s2})}{\alpha_t\,\pi\,B\,D_p} = \frac{360\,P_1}{\alpha_t\,\pi\,B\,D_p}\left[\frac{e^{\mu\alpha_t} + 1}{e^{\mu\alpha_t}}\right] \qquad (11.7)$$

The allowable pressure on pulley surface should be 0·2 to 0·3 MPa for rubber-fabric belts and 0·35 to 0·55 MPa for rubber-cable belts.

11.4.4 Loading devices

Nature of material to be conveyed and method of their delivery on belt should be considered while designing the loading devices. Bags or packets can be loaded on belt directly manually or over a chute. Grain or powder in bulk can be delivered through the hoppers or guide chutes (Fig. 11.1). The height of fall should be as minimum as possible to avoid belt damage and enhance its service life. Moreover, the speed and direction of the material should be close to those of the belt to ensure the smooth delivery. The walls of the hoppers may be cladded with wear-resistant combed rubber. Usually, the length of chute l_{ch} = (1·25 to 2·0) B, h_{ch} = (0·3 to 0·5) B, B_1 = 0·5 B and B_2 = 0·6 B. Powdery materials in bulk should be delivered through hermetically sealed hoppers.

11.4.5 Transfer points

The material on conveyer belt is unloaded from the terminal pulley or at intermediate points by means of a plough and pulley unloaders; called transfer points the latter are employed for bulk materials only. At the terminal pulley (Fig. 11.8), the material fly off the belt and move along a parabola which is described by the coordinates.

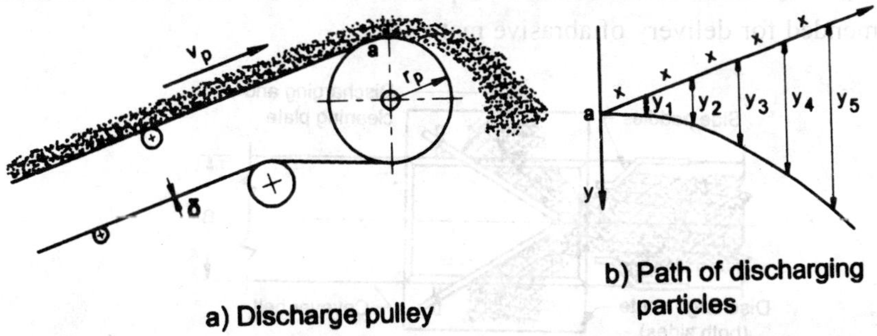

Fig. 11.8 *Motion path of particle discharging from the terminal pulley*

$$x = v_p t \tag{11.8}$$

and $$y = 0{\cdot}5\, g t^2 = \frac{g x^2}{(2 v_p^2)} \tag{11.9}$$

Unloading of material takes place either from a terminal pulley or a tripler at an intermediate point. The latter consists of two pulleys and chutes mounted on a carriage so as to move the unloading point along the whole front (L_f). The material is delivered from upper pulley (Fig. 11.9) to the chute and directed to either one side or both the sides of the belt as desired. These unloaders have the merits that a complete automation is possible and unloading of bulk materials can be done over a long distance. The demerits are their complex design, heavy structure and double turning of belt shortening its service life.

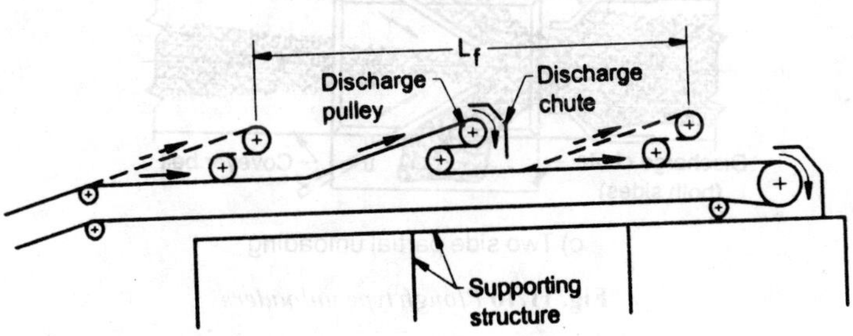

Fig. 11.9 *Pulley type unloader*

A plough-type transfer point consists of a discharging and cleaning plates arranged at 30-40° angle to the longitudinal axis of the belt, receiving hopper mounted on a table, lifting mechanism and side paddles [Fig. 11.10 (*a*)]. The discharging and cleaning plates divert the material on belt either on two sides [Fig. 11.10 (*a*)] or on one side [Fig. 11.10 (*b*)] of the belt since they are in contact with the belt surface. It is also possible to have a diversion of partial material as per requirement, the rest being carried straight on the belt by providing the gap between touching ends of two discharging plates [Fig. 11.10 (*c*)] or by partially covering the belt [Fig. 11.10 (*b*). Plough-type unloaders are used on horizontal conveyers and for 400-2000 mm belt widths unloading the dusty granular and small lump materials of low moisture content with belt speed less than 2.0 m/s. These unloaders are not recommended for delivery of abrasive materials.

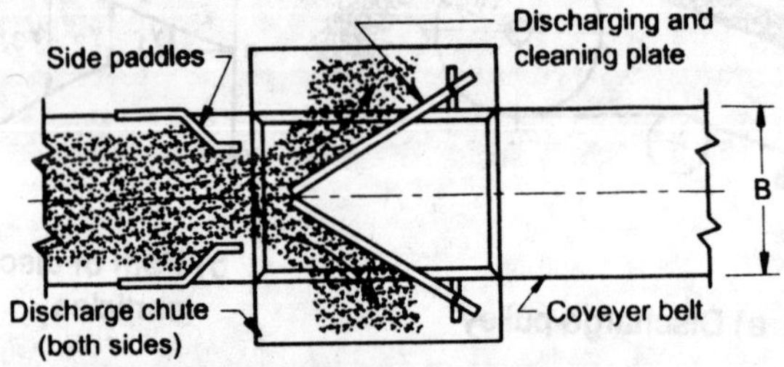

a) Two side unloading

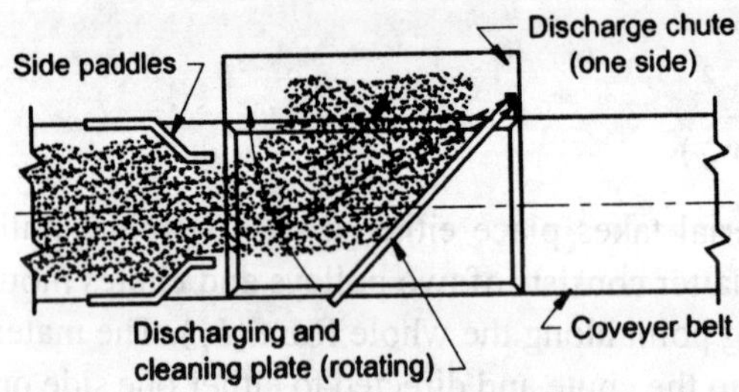

b) One side unloading

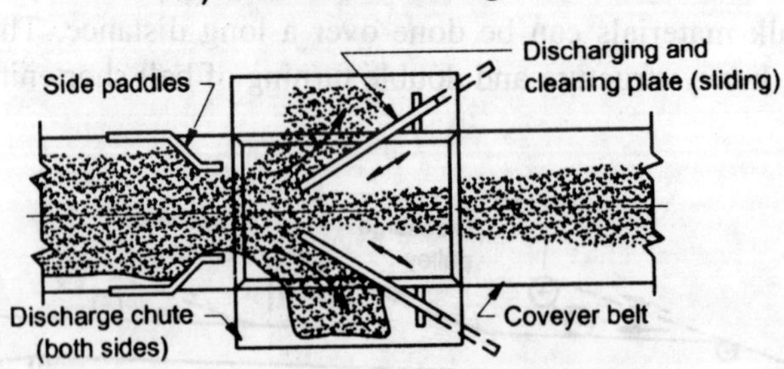

c) Two side partial unloading

Fig. 11.10 *Plough type unloaders*

11.4.6 Take-up devices

Take-up device is an essential part of a tractive element which serves to form the initial tension of the belt, minimizes sag of the belt and compensates for the belt stretching during operation. The initial tension in friction drive such as conveyer belt should be sufficient for normal operation. According to their design and principle of operation, take-up devices are divided as screw type, gravity type, spring type and hydraulic type (Fig. 11.11). Screw type take-ups [Fig. 11.11 (*a*)] are used in horizontal or inclined conveyers up to 60 m long. They are manually operated to adjust the belt tension. A gravity-type take-up [Fig. 11.11 (*b*)] device acts under high hanging weight and automatically maintains a constant tension in the belt, compensating for temperature and reducing peak loads at occasional overstressing. However, the drawbacks are that they are heavy and bulky for powerful and long belt conveyers, employed usually for 60 to 500 m distance. The other types are rarely used.

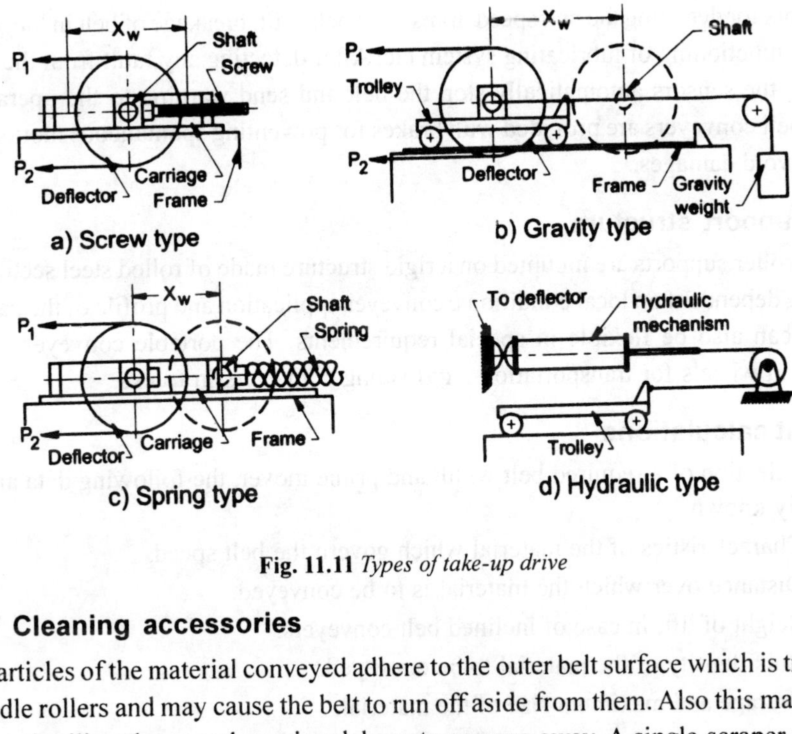

Fig. 11.11 *Types of take-up drive*

11.4.7 Cleaning accessories

Particles of the material conveyed adhere to the outer belt surface which is transferred to the idle rollers and may cause the belt to run off aside from them. Also this material falls on ground, piling there, and requires labour to remove away. A single scraper plate type cleaner [Fig. 11.12 (*a*)] is mostly provided for dry or moist material from load carrying surface of the belt. Wet and sticky materials are cleaned off by a rotating brush [Fig. 11.12 (*b*). The working elements of cleaners are made of wear-resistant rubber, plastic, carbon fibres which do not cause an appreciable wear to the belt surface. Usually, the cleaners are mounted after the belt has delivered the material at head pulley. The cleaned material falls into a hopper.

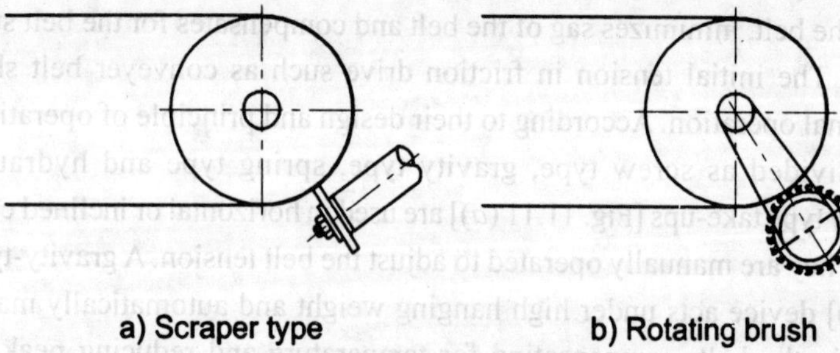

Fig. 11.12 *Types of belt cleaner*

11.4.8 Safety and controls

The modern conveyers are provided with the safety devices and automatic control with sensors for detecting the belt speed, transverse belt shift, breakage of belt in longitudinal direction, functioning of lubricating system etc. After detecting any fault from the normal operation, the sensors automatically stop the belt and send warning to the operator. An inclined belt conveyers are provided with brakes for preventing spontaneous motion of the belt and avoid damages.

11.4.9 Support structure

The roller supports are mounted on a rigid structure made of rolled steel sections, the type being dependent on local conditions, conveyer application and profile of the path. The structure can also be flexible in special requirements. The portable conveyer frame is mounted on wheels for transportation *e.g.* loading conveyers in trucks.

11.5 Belt calculations

For selection of a required belt width and prime mover, the following data are to be specifically known.

(*a*) Characteristics of the material which govern the belt speed.

(*b*) Distance over which the material is to be conveyed.

(*c*) Height of lift, in case of inclined belt conveyers.

(*d*) Bulk density of the material to be conveyed.

(*e*) Average and maximum size of the material.

(*f*) Temperature of the material, if it is hot.

(*g*) Average and maximum load carrying capacity of the belt

(*h*) Any other special requirements.

After the calculations, a diagram of designed conveyer should be drawn with its principal dimensions and data on the expected operating conditions, arrangement and operation of the conveyer and methods of its loading and unloading.

The width of the conveyer belt is determined by the maximum rated capacity of the loading device. On flat or straight rollers [Fig. 11.13 (*a*)], the materials assume an equilateral triangle. The area of cross-section of the material is given by,

$$A = \frac{b\,h}{2}$$

where, $b = (0{\cdot}9\ B - 0{\cdot}05)\ h = 0{\cdot}5\ b \tan \phi_d$

and $A = 0{\cdot}25\ b^2 \tan \phi_d$ (11.10)

Usually, $\phi_d = 0{\cdot}35\ \phi_s$ and for most of the bulk grain $\phi_s = 28$ to $32°$. With the substitution of these values the following equation is obtained.

$$A = 0{\cdot}049\ (0{\cdot}9\ B - 0{\cdot}05)^2 \tag{11.11}$$

Then, the theoretical capacity of the conveyer belt is given by,

$$Q_v = 3600\ A\ v \tag{11.12}$$

Multiplying this capacity by the bulk density of the material, the mass capacity is obtained,

$$Q_t = 3600\ \rho\ A\ v \tag{11.13}$$

From equation (11.11) and (11.13),

$$Q_t = 3600\ [0{\cdot}049\ (0{\cdot}9\ B - 0{\cdot}05)^2]\ v\ \rho \tag{11.14}$$

The shape of material (cross-section) on troughed belt [Fig. 11.13 (*b*)] depends on the troughing angle β. It is the sum of the equilateral triangle due to dynamic angle of repose and the lower trapezium whose sides are formed by the side roller supports. The total area for β = 20° is given by,

$$A_{20} = 0{\cdot}097\ (0{\cdot}9\ B - 0{\cdot}05)^2 \tag{11.15}$$

and $Q_{t\,20} = 350\ (0{\cdot}9\ B - 0{\cdot}05)^2\ v\ \rho$ (11.16)

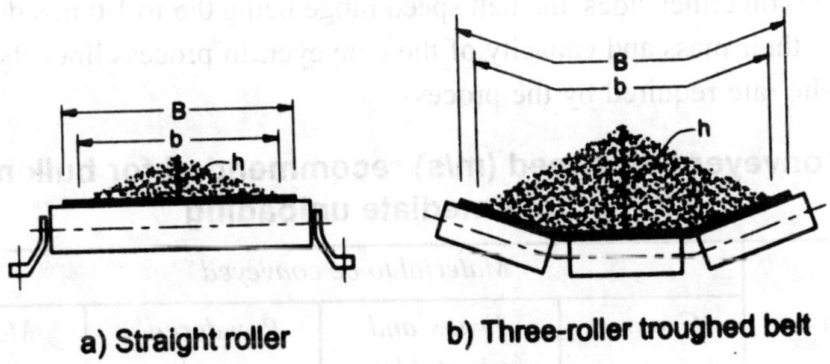

Fig. 11.13 *Load section on conveyer belt*

For β = 30°, the cross-sectional area and capacity are given by,

$$A_{30} = 0{\cdot}124\ (0{\cdot}9\ B - 0{\cdot}05)^2 \tag{11.17}$$

and $Q_{t\,30} = 445\ (0{\cdot}9\ B - 0{\cdot}05)^2\ v\ \rho$ (11.18)

From equation (11.16) and (11.18), it is obvious that the conveyer belt capacity increases with the increase in troughing angle but its life is decreased. The troughing angle $\beta = 20°$ is found to be the most economical. The required belt width is determined by the following expression,

$$B = 1{\cdot}1\ b + 0{\cdot}056 \tag{11.19}$$

In case of inclined belt conveyers, the cross-sectional area of the material is reduced resulting in less capacity which is expressed as,

$$Q_r = C_r \cdot Q_t \tag{11.20}$$

The coefficient of reduction of load layer cross-section on an inclined portion of the belt conveyer, is also called as the fall back factor, the values of which are taken as below,

θ, deg	:	< 2	4	6	8	10	12	14	16
C_r	:	1·00	0·99	0·98	0·97	0·95	0·93	0·91	0·87
θ, deg	:	18	20	22	24	26	28	30	
C_r	:	0·85	0·81	0·76	0·71	0·66	0·61	0·56	

11.6 Selection of belt speed

The selection of conveyer belt speed can be done from Table 11.4, depending on the nature of material to be conveyed, belt width, number of unloading stations, application and location of the conveyer. The belt speed should not exceed 2·0 m/s for pulley type unloaders and for plough type 1·0 to 1·6 m/s. Wide belts can have higher speed than narrower because of less fall out of the material. Indoor belts have lower speed than outdoor ones. For inclined conveyers lower value from the speed range should be chosen so as to avoid material spilling. It is necessary for the unit load conveyers to provide 50 to 100 mm free edges on either sides, the belt speed range being 0·5 to 1·0 m/s depending on the type of units, their mass and capacity of the conveyer. In process lines the belt speed is governed by the rate required by the process.

Table 11.4 Conveyer belt speed (m/s) recommended for bulk materials without intermediate unloading

Belt width, mm	*Material to be conveyed*			
	Grain	*Fruits and vegetables*	*Powdered/ granular material*	*Medium lumpy material*
400 - 500	1.6	0.8	1.3 - 1.6	1.3 - 1.6
650 - 800	2.0 - 2.5	0.8	2.0 - 2.5	1.6 - 2.0
1000 - 1200	3.2 - 4.0	1.0	2.5 - 4.0	2.0 - 2.5
1400 - 1600	—	—	3.2 - 4.0	2.5 - 3.1

11.7 Determination of belt tension

For the computation of tension in belt, which is equal to the tractive force exerted by the drive, the following generalized equation is used.

$$P = f_p\, L_h\, \{(q_m + q_c + q_b)\, f_c + (q_b + q_r)\, f_r\} \pm q_m\, H \tag{11.21}$$

The coefficient of resistance at pulleys, unloading stations and other places is based on the belt length as below.

Belt length, m	:	10	20	30	40	50	100	200	300	500	>1000
f_p	:	4·5	3·2	2·8	2·6	2·4	1·7	1·5	1·4	1·3	1·1

For flat steel or wooden rollers $f_c = 0{\cdot}35$ to $0{\cdot}60$ depending on the operating conditions; f_c for three-roller troughed belt has the following values.

Condition of operation	:	Good	Moderate	Heavy (summer)	Heavy (winter)
Conveyer length, m					
up to 100	:	0·020	0·025	0·035	0·045
above 100	:	0·018	0·022	0·032	0·042

The coefficient of resistance to motion for the return run and is equal to 0·018, 0·022, 0·030 and 0·040, respectively for the above conditions. The linear gravity forces of different elements can be added to the equation (11.21) resulting as below for bulk loads.

$$P = f_p\, L_h\, g \left\{ \left(\frac{Q_t}{3{\cdot}6v} + \frac{m_c}{l_c} + m_b \right) f_c + \left(m_b + \frac{m_r}{l_r} \right) f_r \right\} \pm q_m H \tag{11.22}$$

For the unit loads l_c is replaced by l_p is above equation. Plus sign is taken for up-slope and minus sign for down slope conveying.

Alternately, the total pull is roughly determined by the formula,

$$P = C_m\, L_b\, f\, g\, (w_{br} + w_m) \pm w_m g H \tag{11.23}$$

Factor which takes into account all the resistances to belt motion and its value depends on the belt length L_b as below.

L_b, m	:	3	4	5	6	8	10	12·5	16	20	25	32	40
C_m	:	9·0	7·6	6·6	5·9	5·1	4·5	4·0	3·6	3·2	2·9	2·6	2·4
L_b, m	:	50	63	80	100	125	200	250	320	400	500	> 600	
C_m	:	2·2	2·0	1·85	1·7	1·6	1·4	1·3	1·2	1·1	1·05	1·00	

The coefficient of friction of idler bearings is taken as 0·025 for antifriction and 0·050 for slip (with good lubrication) bearings.

Power requirement on the drive shaft is computed by the following equation.

$$P_{ds} = \frac{P v}{986} \tag{11.24}$$

and that at motor as,

$$P_m = \frac{P_{ds}}{(\eta_d\, \eta_t)} \text{ kW} \tag{11.25}$$

Example 11.1 : *Design a 60 t/h three-roller 20° troughed belt conveyer to carry wheat grain to be stored in silos up to a 57·5 m horizontal distance. The weight of belt, rollers on carrying and return runs may be taken as 16 kg/m, 50 kg and 24 kg, respectively. The grain is to be delivered in the silos on to the terminal pulley where angle of belt wrap is 166°. Power input to the belt shall be through a double-drum drive for which provide an angle of wrap of 174° and 220°, respectively for first and second drums.*

Solution.

Given, Q_t = 60 t/h, β = 20° for three-roller system, L_h = 57·5 m, w_b = 16 kg/m, w_c = 50 kg, W_r = 24 kg, α = 166° = 2·90 rad, α_1 = 174° = 3·03 rad, α_2 = 220° = 3.84 rad

Assume bulk density for wheat grain, ρ = 0·82 t/m³, angle of repose, ϕ = 30°, then the dynamic angle of repose, ϕ_d = 0·35 × 30 = 10·5, belt velocity, v = 1·6 m/s.

Then $Q_t = 350\, b^2\, v\, \rho$

or $\quad 60 = 350\, b^2 \times 1{\cdot}6 \times 0{\cdot}82$

then $\quad b = 0{\cdot}36$ m

and $\quad B = 1{\cdot}1 \times 0{\cdot}36 + 0{\cdot}05$

$\quad = 0{\cdot}446$ m

Hence, 450 mm belt can be adopted. The capacity for this belt from equation (11.16) at recommended speed, v = 1·6 m/s (Table 11.4) will be,

$$Q_t = 350\,(0{\cdot}9 \times 0{\cdot}45 - 0{\cdot}05)^2 \times 1{\cdot}6 \times 0{\cdot}82$$

$$= 57{\cdot}87 \text{ t/h}$$

The belt will be overloaded for a given capacity, so the next higher belt width of 500 mm can be provided. Then, Q_t = 73·47 t/h which is about 22% higher than given capacity.

The roller pitch for carrying run can be adopted as l_c = 1·4 m (Table 11.2) and l_r = 3.0 m for return run as specified. Also f_p = 2·3, f_c = 0·035, f_r = 0·030 and H = 0. Then, using equation (11.22),

$$P = 2\cdot3\times57\cdot5\times9\cdot81\left[\left(\frac{60}{3\cdot6\times1\cdot6}+\frac{50}{1\cdot4}+16\right)\times0\cdot035+\left(16+\frac{24}{3\cdot0}\right)\times0\cdot030\right]\pm0$$

or $\quad P = 3755{\cdot}4$ N

Power requirement at the drive shaft will be,

$$P_{ds} = \frac{3755\cdot4\times1\cdot6}{986} = 6\cdot09 \text{ kW}$$

Considering the drive assembly, transmission and prime mover efficiency, the total power requirement will be,

$$P_m = \frac{6\cdot09}{(0\cdot85\times0\cdot80\times0\cdot90)} = 9\cdot95 \text{ kW}$$

Hence, provide 11 kW (15 horse power) electric motor as a prime mover for the belt conveyer. Then, total pull is,

$$P = \frac{736\, P_m}{v} = \frac{736\,(15 \times 0{\cdot}85 \times 0{\cdot}80 \times 0{\cdot}90)}{1{\cdot}6} = 4222{\cdot}80 \text{ N}$$

but $P = P_1 - P_3 = 4222{\cdot}80$

and $P_1 = P_2\, e^{\mu\alpha} = P_2\, e^{0{\cdot}35 \times 2{\cdot}9} = 2{\cdot}76\, P_2$

For double-drum drive, $\alpha_t = \alpha_1 + \alpha_2 = 3{\cdot}04 + 3{\cdot}84 = 6{\cdot}88$ rad

also $P_2 = P_2\, e^{\mu\alpha} = P_3\, e^{0{\cdot}35 \times 6{\cdot}88} = 11{\cdot}11\, P_3$

then $P_1 = 2{\cdot}76\, P_2 = 2{\cdot}76\,(11{\cdot}11\, P_3) = 30{\cdot}66\, P_3$

Thus $30{\cdot}55\, P_3 - P_3 = 4222{\cdot}80$

then $P_3 = 142{\cdot}40$ N

$P_2 = 11{\cdot}11 \times 142{\cdot}40 = 1582{\cdot}1$ N

and $P_1 = 30{\cdot}66 \times 142{\cdot}40 = 4366$ N

Assuming four-ply belt, different pulley diameters are,

$D_p = 125 \times 4 = 500$ mm

$D_r = 100 \times 4 = 400$ mm

$D_g = 70 \times 4 = 280$ mm

Actual belt pressure on pulley from equation (11.7) is

$$\sigma_b = \frac{360 \times 4366}{6{\cdot}88\pi \times 500 \times 500}\left[\frac{11{\cdot}11 + 1}{11{\cdot}11}\right] = 0{\cdot}317 \text{ N/mm}^2$$

Since, the actual belt pressure on drive pulley is very small as compared with the allowable pressure, the provision of belt and pulley is alright.

Provide spacing of the supporting rollers on carrying side as 1·40 m (Table 11.2) and 3·0 m on return run of the conveyor belt. All the troughing and supporting rollers shall be 89 mm diameter (Table 11.1).

Nomenclature

A : area of cross section of material on belt, m^2

A_{20} : area of cross section of material on belt with 20° troughing angle, m^2

A_{30} : area of cross section of material on belt with 30° troughing angle, m^2

B : width of the belt, mm

b : width of material on belt, mm

C_d : coefficient multiplier for determination of drive pulley diameter

C_f : factor of safety for belt plies

C_m : factor considering all the resistances to belt motion

C_r : coefficient of load reduction on inclined belt

D_g : diameter of guide pulley, mm

D_p : diameter of drive pulley, mm

D_r : diameter of return pulley, mm

f : coefficient of friction of idler bearing
f_c : coefficient of resistance to load carrying belt motion
f_p : coefficient of resistance at pulleys
f_r : coefficient of resistance to return run belt motion
g : gravitational constant, m/s^2
H : vertical lift height of material on inclined belt, m
h : maximum height of material on belt, m
h_{ch} : height of the chute, m
L_b : length of the belt, m
L_f : length of tripler travel for material unloading, m
L_h : horizontal projected distance between terminal pulleys, m
l_c : roller spacing on carrying side of belt, m
l_{ch} : length of the chute, m
l_p : spacing between adjacent pieces of unit load on carrying side of belt, m
l_r : roller spacing on return run of belt, m
m_b : mass of the belt per unit length, kg/m
m_c : total mass of rollers on carrying side of belt, m
m_r : total mass of rollers on return run of belt, m
N : number of plies in belt
P : total pull (tension) in belt, N
P_{ds} : power requirement at drive shaft, kW
P_m : power requirement for electric motor, kW
P_{max} : maximum rated pull (tension) in belt, N
P_1 : pull (tension) on tight side of belt, N
P_2 : pull (tension) on slack side of belt on first or single pulley, N
P_3 : pull (tension) on slack side of belt on second pulley, N
P_{S1} : pull (tension) on tight side of belt in steady motion, N
P_{S2} : pull (tension) on slack side of belt in steady motion, N
Q_r : reduced theoretical capacity of belt, t/h
Q_v : volumetric theoretical capacity of belt, m^3/h
Q_t : theoretical capacity of belt, t/h
Q_{t20} : theoretical capacity of belt at 20° troughing angle, t/h
Q_{t30} : theoretical capacity of belt at 30° troughing angle, t/h
q_b : linear gravity forces due to belt weight, N/m
q_c : linear gravity forces due carrying side rollers weight, N/m
q_m : linear gravity forces due to material weight to be conveyed, N/m
q_r : linear gravity forces due to return run rollers weight, N/m

t	: time of particle motion during unloading at terminal pulley, s
v	: linear velocity of conveyer belt, m/s
v_p	: velocity of discharging particle at terminal pulley, m/s
w_{br}	: weight of belt and rollers on carrying side and return run, kg/m
w_m	: weight of material to be carried on belt, kg/m
X_w	: length of take up adjustment, m
x	: distance traveled by the particle along x-axis, m
y	: distance traveled by the particle along y-axis, m
α	: wrapping angle of belt on pulley, deg
α_1	: wrapping angle of belt on first pulley, deg
α_2	: wrapping angle of belt on second pulley, deg
α_t	: total wrapping angle of belt on first and second pulleys, deg
β	: troughting angle of belt, deg
θ	: angle of inclination of belt conveyer with horizontal, deg
η_d	: efficiency of the drive shaft, %
η_t	: efficiency of the transmission system, %
ϕ_d	: dynamic angle of repose of the material, deg
ϕ_s	: static angle of repose of the material, deg
μ	: coefficient of friction between belt and pulley surfaces
μ_1	: coefficient of friction between belt and first pulley surfaces
μ_2	: coefficient of friction between belt and second pulley surfaces
ρ	: bulk density of the material to be conveyed, t/m^3
σ_a	: allowable belt pressure on pulley surface, N/mm^2
σ_b	: belt pressure acting on pulley surface, N/mm^2
σ_n	: ply fabric strength per unit width of belt, N/mm^2 mm

12

Bucket Elevators

12.1 Introduction

Elevators are the conveying machines for handling of bulk materials or unit loads on a vertical or steep slope (60-82° to the horizontal). Bucket elevators are very efficient devices and are adapted variedly and range from simple small capacity unit used in the grain cleaning equipment to the large, expensive units used for granular, pulverized and lumpy materials in large industrial plants and building constructions. They lift the loads from initial to the final point, without any intermediate loading or unloading the material. Bucket elevators are divided into vertical and inclined according to their path of lifting. They are also divided into stationary and mobile, transport and process elevators.

12.2 Construction

A bucket elevator (Fig. 12.1) essentially consists of a vertical tractive element (belt or chain) with load carrying elements (buckets) fixed rigidly to it by rivets or nut-bolts and enclosed in a single housing called a leg, or two legs may be used separately for carrying and return runs. A tractive element is an endless running on upper drive pulley and lower take-up pulley for belts and in case of chain, sprockets are used. The top portion of casing above legs is called as head and the bottom portion as boot. Buckets are fastened to the tractive element at the back and the later is powered by an electric motor through the drive. An initial tension is set up by a screw type or spring loaded take-up. Bulk material is fed at the boot into the hopper where flow rate is controlled by a flap (sliding) valve. It is then poured into the buckets, lifted up and discharged at the head through a spout. A stop or break is provided to the drive to prevent backward motion of the tractive element at the end of work or meanwhile failure of power. Guides are sometimes provided for two-chain tractive element, particularly in the up leg. Single chain and belt installations have no guides, but one or two idlers may be provided to eliminate whip.

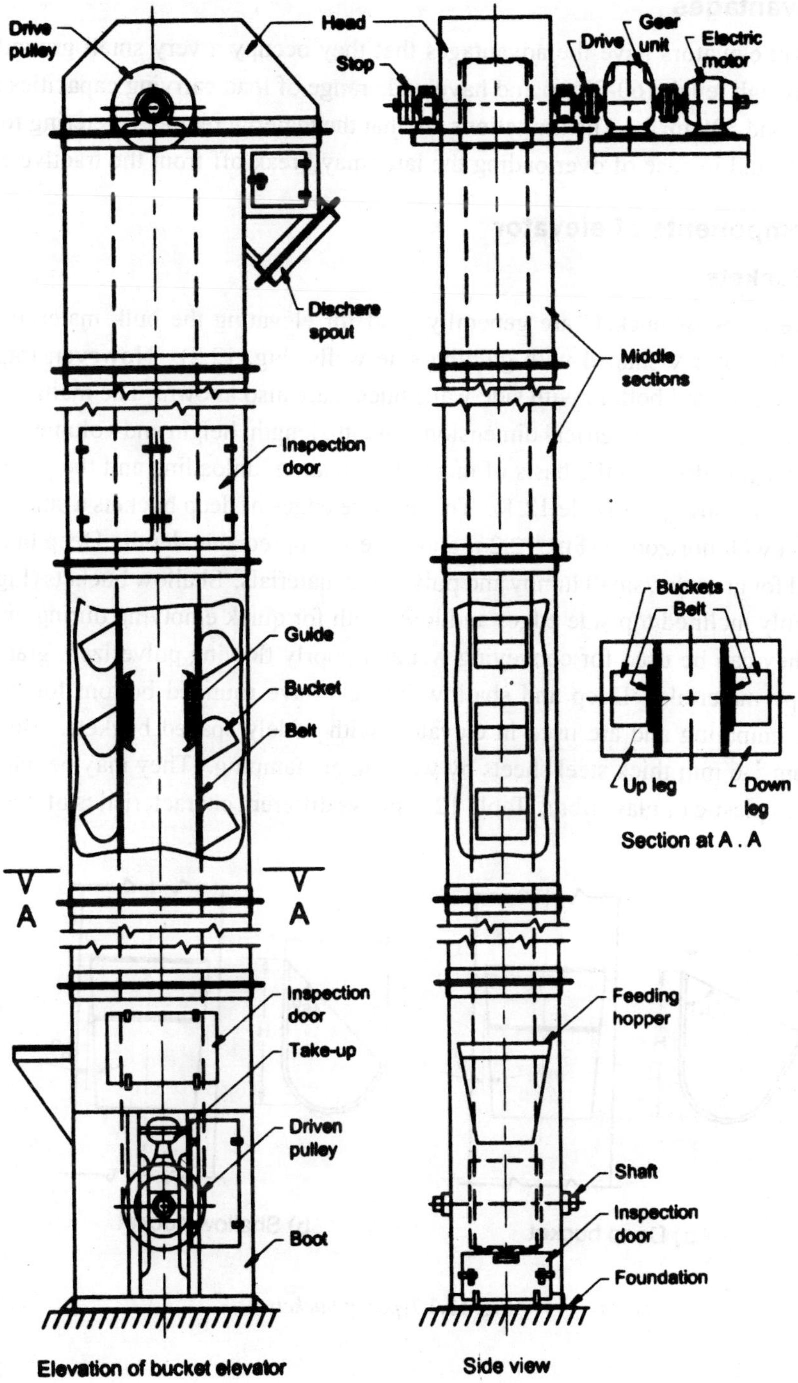

Fig. 12.1 *Bucket elevator assembly*

12.3 Advantages

Bucket elevators have the advantages that they occupy a very small ground space, lift load to high levels (60-75 m) and have wide range of load carrying capacities ranging between 5 and 500 m^3/h. The limitations are that the material requires metering for filling the buckets and in case of overloading the later may break off from the tractive element.

12.4 Components of elevator

12.4.1 Buckets

Three types of buckets are generally used for elevating the bulk material namely, deep, shallow and V-shaped with guiding side walls (Fig. 12.2). However, trapezoidal buckets and rounded bottom with side walls bucket are also known. The main features of the buckets are their geometrical dimensions : width, length; height and volume. Selection of the buckets is done on the basis of method of loading, unloading and the properties of material to be conveyed (Table 12.1). The top side edges of deep buckets a small angle of inclination with horizontal (Fig. 12.2*a*) and have an appreciable depth. Deep buckets are employed for granular, small lumpy and pulverized materials. Shallow buckets (Fig. 12.2*b*) have steeply inclined top side edges and low depth for quick emptying during unloading. Hence, they can be used for conveying wet and poorly flowing pulverized, granular and fine lumpy materials. Deep and shallow buckets have rounded bottom for better and complete emptying and are used in elevators with widely spaced buckets. Buckets are made from 1-6 mm thick steel sheets by welding or stamping. They may be made of pig iron, fibres plastic or glass fibre. Table 12.2 shows different characteristics of the standard buckets.

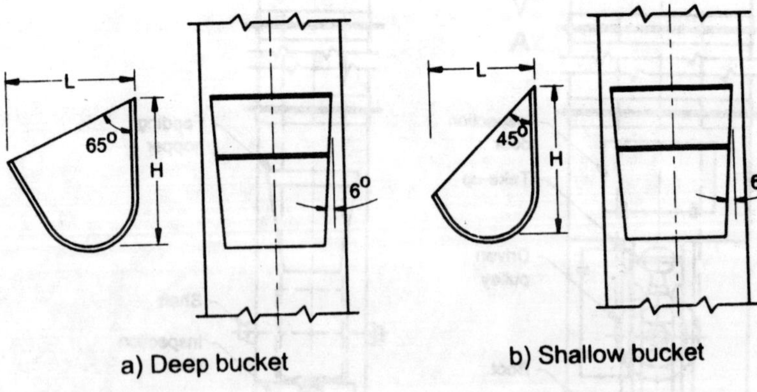

a) Deep bucket b) Shallow bucket

Fig. 12.2 *Types of buckets*

Table 12.1 Selection of bucket

Material to be conveyed	*Type of elevator*		*Bucket type*	*Bucket speed, m/s*	*Coefficient of filling (ψ)*
	Speed	*Unloading*			
Food grain	high	centrifugal	deep	2·0 - 3·2	0·75
Grain products (flour, powder)	high	centrifugal, gravity	shallow	1·0 - 1·6	0·85
Sand, soil, ash	high	centrifugal	deep	1·0 - 2·0	0·80
Saw dust, fine coal	low	gravity	deep	1·3 - 2·0	0·80

Table 12.2 Specification for widely spaced buckets and belts

Width of bucket, mm	*Belt width, mm*	*Bucket pitch, mm*	*Bucket volume, lit*		*Volume/pitch ratio, lit/m*	
			Deep	*Shallow*	*Deep*	*Shallow*
100	125	200	0·2	0·10	1·0	0·50
125	160	220	0·4	0·20	1·3	0·66
160	200	320	0·6	0·35	2·0	1·17
200	250	400	1·3	0·75	3·2	1·87
250	300	400	2·0	1·40	5·0	3·50
320	370	500	4·0	2·70	8·0	5·40
400	450	500	6·3	4·20	12·6	8·40
500	550	630	12·0	6·80	19·0	10·80
650	700	630	16·8	11·50	26·6	18·20

12.4.2 Tractive element

A tractive element may be a belt similar to those used in belt conveyers or a chain or two. A hollow hole is provided in the back of buckets for fastening to the belt [Fig. 12.3 (*a*)] and leaving a flat surface of belt with bolt to pass on the drive pulley, or bucket may be fixed with plane surface. A belt must be 25 to 100 mm wider than the bucket width with at least four plies [Fig. 12.3 (*b*)]. For heavy work, the number of plies is determined from the effective pull. Generally, laminated, sieve type chains are used for elevators with 100 to 630 mm pitch or welded round-link chains from 16 to 28 mm diameter steel rods with heat treatment for reducing the wear. An angle or shaped sections are used to fasten up to 250 mm wide buckets on chain. For wider buckets, two chains can be used. Belts are used for the conveyance of abrasive materials at higher speeds whereas chains are employed in elevators of large capacity at a substantial height and hot materials.

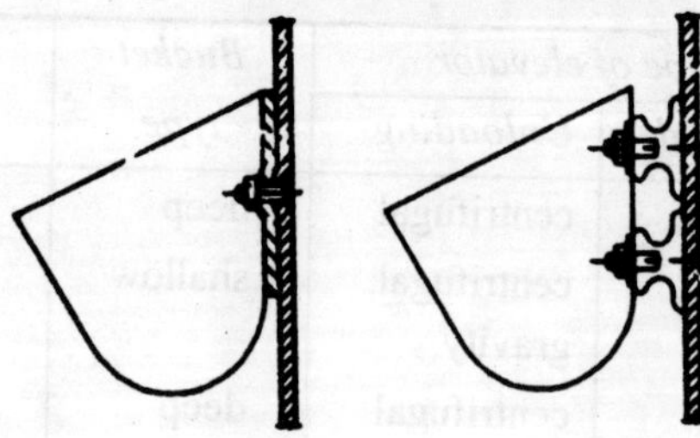

Fig. 12.3 *Fastening of bucket on belt*

12.4.3 Drive

Usually, bucket elevators are driven by a geared electric motor installed in its head. A method of unloading the buckets decides the diameter of a drive pulley. The criterion of number of plies should also be considered for pulley diameter [usually, D_p = (125 to 150) N, mm]. With this determined diameter the standard size should be adapted. For enhancement of the coefficient of friction with the belt, rubber-lined belt pulley can be used. The drive sprocket with 5 to 20 teeth is used for laminated chains.

12.4.4 Take-up

In order to keep the belt or chain straight from upper and lower pulleys, a screw, spring or gravity type take-ups are employed (Fig. 12.1) on the shaft of lower pulley and fastened to the side walls of elevator boot. The take-up travel is about 200 to 500 mm.

12.4.5 Casing

It is a mid-section of a bucket elevator 2·0-2·5 m high made of mild steel sheets 2·4 mm thick. The head and boot sections are bolted with middle section by means of bolts with sealing washers. The casing is a supporting structure of a bucket elevator and absorbs static and dynamic forces. The inlet hopper is provided in the elevator boot which may be high with bottom inclined at 60° with horizontal used for wet poorly flowing materials, or low with 45° inclination for free-flowing or granular materials. A flap valve, sliding up and down, is provided to control the feed rate into the buckets. An inspection door is also provided at the middle section.

12.5 Methods of bucket loading

The material may be loaded in buckets either by scooping it at boot [Fig. 12.4 (*a*)] or by pouring the material into the buckets [Fig. 12.4 (*b*)]. Scooping is usually associated with widely spaced buckets for conveying dry powdery and granular materials which can be easily scooped at 0·8 to 4·0 m/s belt speed. Large-size lumpy and abrasive materials are mostly poured directly into the buckets which are closely placed and running with the speed lower than 1·0 m/s. At higher speeds the material may be thrown off leaving the bucket incompletely filled. In practice, elevator buckets are filled by both methods, one of them being predominant.

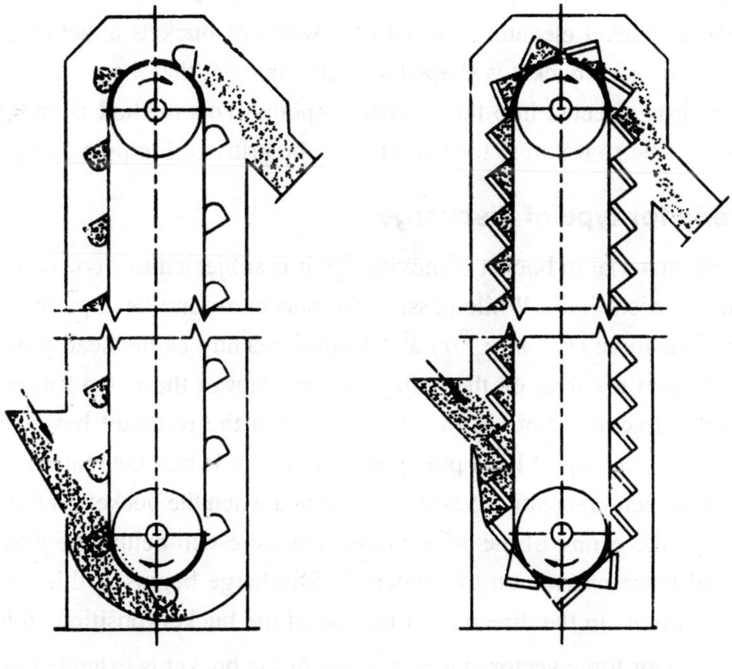

(*a*) Scooping buckets and centrifugal discharge

(*b*) Direct tilling and gravity discharge

Fig. 12.4 *Loading and unloading of buckets*

12.6 Methods of bucket unloading

Normally, three types of bucket unloading are used in bucket elevators *viz*., centrifugal discharge, positive or perfect or gravity discharge and continuous or directed gravity discharge. In case of centrifugal discharge, the material is unloaded by a centrifugal force action as the buckets pass over the head pulley [Fig. 12.4 (*a*)] into spout of an elevator. It is employed, generally in highspeed belt (1 to 4 m/s) elevators for free-flowing, pulverized or granular and small to medium size lumps with widely spaced buckets. The choice of belt speed is critical so that either the material should not fall-off or should not strike the preceding bucket. They are generally loaded by scooping-off the material at the bottom of boot at the foot of the elevator.

Positive or perfect or gravity discharge is achieved by providing a snub roller or guide bar just after the head pulley. This causes the buckets deflected towards the centre of an elevator [Fig. 12.4 (*b*) leaving sufficient space for location of a discharge spout below the head pulley so that the discharge by gravity can be performed. Loading is by scooping the material at the boot. As this type of discharge does not require high speed (<1·0 m/s), it is generally employed for handling of light, fluffy, dusty or some sticky materials that will not be easily discharged by the centrifugal forces.

Continuous bucket elevators consist of a series of buckets attached to a belt [Fig. 12.4 (*b*)]. Front of each bucket is shaped in such a manner that the material poured from a bucket onto it is directed into the discharge spout. This method is employed in low-speed elevators (0·4 to 0·8 m/s) for handling of the gritty or lumpy materials.

12.7 Criteria for type of discharge

When the material in bucket is moving up, it is subjected to a constant gravity force ($P = mg$) in a vertical line. While passing the bucket around head pulley, an additional radial centrifugal force ($F = m\,v_m^2/r$) acts from the centre of the head pulley, [Fig. 12.5 (*a*)]. As the bucket revolves on the pulley, the resultant of these two forces varies in the magnitude and direction. Note in Fig. 12.5 (*b*), that the resultant has such a direction before reaching to the top of head pulley (position 1 to 4) that the material is held in the bucket. It will be zero at position 5 which is attained when the bucket is at the vertical line passing through the centre of the drive pulley and there will neither be gravity force nor the centrifugal force exerted on the material. Discharge begins at this point since, the resultant force points in the direction of motion of the bucket (positions 6 to 8).

If the resultant force vector at any position of the bucket is extended to intersect the vertical line passing through the centre of head pulley, it intersects at one and the same point (A) which is called as the pole as shown in Fig. 12.6 (*a*). The height of pole (h) from the centre of head pulley (O) is termed as the pole distance. From the similar triangles AOB and BCD in Fig. 12.6 (*a*),

$$\frac{h}{r} = \frac{P}{F} = \frac{m g}{\left(m v_m^2 / r\right)}$$

which gives the pole distance as,

$$h = \frac{g r^2}{v_m^2} \tag{12.1}$$

but
$$v_m = \frac{2 \pi r n}{60} \tag{12.2}$$

Substituting this in equation (12.1),

$$h = \frac{(60)^2 \, g r^2}{(2 \pi r n)^2}$$

or $$h = \frac{895}{n^2} \tag{12.3}$$

Equation (12.3) clearly shows that the pole distance is a function of rotational speed of the head pulley only. At higher speeds, the pole distance decreases, $h < r_p$ as shown in Fig. 12.6 (*b*), increasing the magnitude of centrifugal force as compared with that of the gravitational force. This causes the material in bucket to fall in the left side leg of the casing, reaching a very small quantity to the discharge chute. On the contrary, at lower speeds, an effect of gravity force on material increases over the centrifugal force and the pole distance increases, $h > r_p$ [Fig. 12.6 (*a*)]. It results in the fall of material in bucket on the back of former bucket and hence, in the right leg of casing instead of reaching the discharge chute. Therefore, an optimum speed of the belt is very necessary for an effective discharge of the conveyed material into the discharge chute, when $r_p < h \leq r_a$ condition exists, the material in bucket is unloaded by combined method (centrifugal and gravity). Hence, the method of bucket unloading is determined by the ratio of pole distance to effective pulley radius, that is,

$$R_p = \frac{h}{r_p} \tag{12.4}$$

From equations (12.1) and (12.4),

$$R_p = \frac{h}{r_p} = \frac{g r^2}{v_m^2 r_p} = \frac{g r_p}{v_p^2} \tag{12.5}$$

Thus, the method of unloading the bucket is decided not by the bucket velocity, but by the ratio of pulley diameter and its peripheral velocity. A centrifugal discharge can be performed at low bucket speed by providing a small pulley diameter and vice versa. The following ratios are recommended by Fadeev N.K. for different belt speeds :

(*a*) High speed elevators with centrifugal discharge ($R_p \leq 1{\cdot}0$),

$$D_p = \frac{2\,R_p v_b^2}{g} \leq 0.204\, v_b^2 \tag{12.6}$$

(*b*) High speed elevators with combined discharge ($R_p = 1{\cdot}0$ to $1{\cdot}4$),

$$D_p = (0{\cdot}205 \text{ to } 0{\cdot}286)\, v_b^2 \tag{12.7}$$

(*c*) Moderate speed elevators with combined discharge ($R_p = 1{\cdot}5$ to $3{\cdot}0$),

$$D_p = (0{\cdot}306 \text{ to } 0{\cdot}612)\, v_b^2 \tag{12.8}$$

(*d*) Low speed elevators with gravity discharge ($R_p > 3{\cdot}0$),

$$D_p = 0{\cdot}6\, v_b^2 \tag{12.9}$$

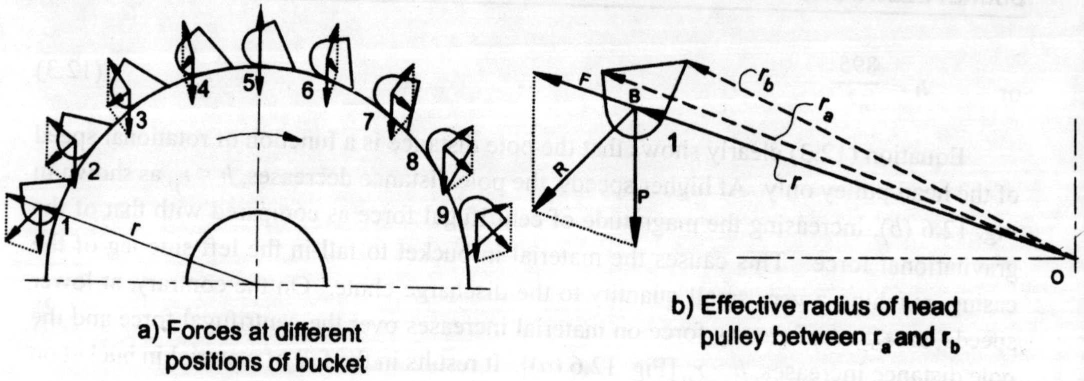

Fig. 12.5 *Force acting on grain in bucket during unloading*

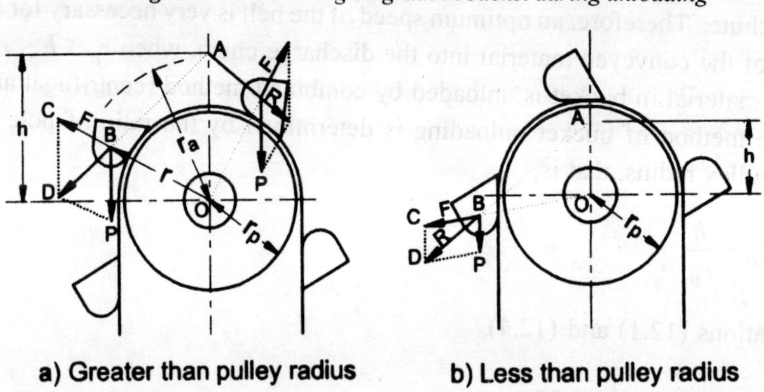

Fig. 12.6 *Pole distance in bucket elevator.*

12.8 Particle trajectory during discharge

It is assumed that the particle leaves the bucket when centrifugal force exceeds the gravity force and travels in a parabolic, path with an absolute velocity equal to the circumferential velocity ($v_c = \omega\, r_a$) at that point neglecting the interaction between the particles. If the sliding of particles are to be considered, the absolute velocity is the geometrical sum of circumferential and sliding velocities [Fig. 12.7 (*a*)].

Beginning of free motion of particles depends on the method of discharge. In case of centrifugal discharge [Fig. 12.7 (*b*)] the particles begin to move inside the bucket as the later has rotated through an angle β = 15 to 30°. The position of the particle (*x*, *y*) at any time *t* which has left the bucket at point (x_o, y_o) *i.e.* ($r \sin \beta$, $r \cos \beta$) at time $t = 0$ can approximately be determined as below :

$$x = r \sin \beta + (\omega\, r \cos b)\, t \qquad (12.10)$$

$$y = r \cos \beta - (\beta\, r \sin b)\, t - \frac{g\, t^2}{2} \qquad (12.11)$$

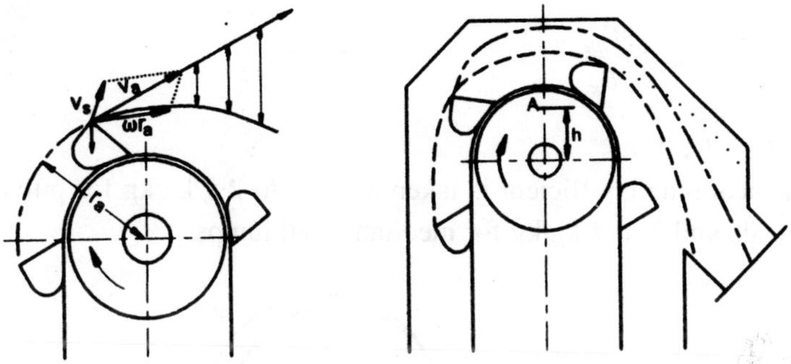

a) Velocity components of the discharging particle

b) Path of particle in the centrifugal discharge

Fig. 12.7 *Trajectory path of particle*

The head casing should be shaped according to the trajectories of particle during discharge from the bucket and situated such that all the ejected particles are directed to the discharge spout and should neither strike the casing wall nor fall back on the return run of the bucket elevator. In former case, excessive breakage of the material and dusting will result whereas, elevator efficiency will be reduced in the later case. The position of a discharge spout is determined by the unloading angle β (Fig. 12.7 *b*), the value of which ranges between 30 and 45° for centrifugal and combined discharge type elevators.

12.9 Elevator calculation

Considering the characteristics of the materials to be handled, the type of bucket elevator and the shape of bucket are selected (Table 12.2). Capacity of an elevator can be determined by using the following equation.

$$Q_t = 3{\cdot}6\ v_b\ \rho\ C_f\ \frac{V}{B_s} \tag{12.12}$$

The coefficient of bucket filling is (taken as 0·4 to 0·5 for big lumpy material, 0·6 to 0·7 for small lumps and coal, 0·75 to 0·90 for granular materials and 0·8 to 1·0 for powdery materials. The bucket spacing (pitch) on the belt is taken as (2·5 to 3·0) H for widely spaced buckets and equal to H for closely spaced buckets. Elevators with closely spaced buckets can ensure a high carrying capacity even at reduced belt speeds, since they have a high V/B_s ratio.

12.10 Belt pull and power requirement

Referring to Fig. 12.8, and considering the geometry,

$$w_m = \frac{Q_t}{(3{\cdot}6 v_b)} \tag{12.13}$$

$$w_b = w_{be} + \frac{W_{bu}}{B_s} \tag{12.14}$$

and $$P_s = C_s\, w_m \cdot g \tag{12.15}$$

The scooping coefficient is taken as 1·25 to 2·5 kg/kg for pulverized and fine size lumpy loads and 2 to 4 kg/kg for medium sized lumps.

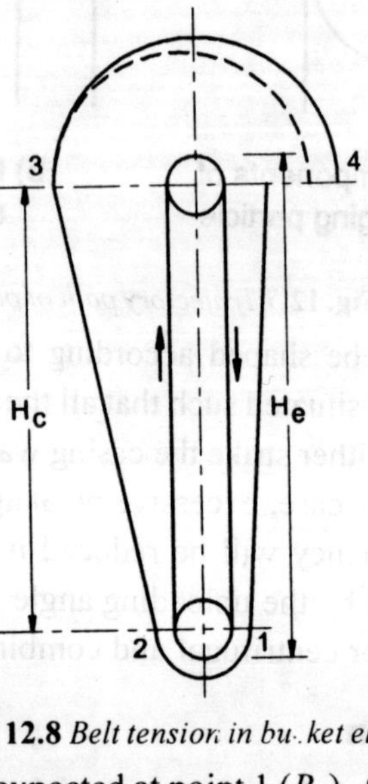

Fig. 12.8 *Belt tension in bucket elevator*

The lowest belt tension is expected at point 1 (P_1). Considering this and the scooping resistance for light and medium operation at $C_s = 2{\cdot}5$, the belt tension at point 2 is given by the following formula :

$$P_2 = P_1 + P_s = P_1 + C_s\, w_m \cdot g \tag{12.16}$$

$$P_3 = P_2 + g\,(w_m\, H_e + w_b\, H_c) \tag{12.17}$$

and $$P_4 = P_1 + w_b . g . H_c \tag{12.18}$$

Also, $$P_3 = P_4\, e^{ma}. \tag{12.19}$$

where e^{ma} is the pull factor which determines the tractive capacity of the driving pulley explained in equation (11.2).

From the above equations the value of P_1 can be determined which, in turn, will yield the magnitude of other belt tensions. The total pull on the head pulley is given by,

$$P_t = P_3 - P_4 \tag{12.20}$$

and the power requirement at drive shaft can be computed using the equation (11.24) at prime mover power from equation (11.25).

***Example 12.1 :** A bucket elevator is to be provided for cleaner-cum-grader having 10 tph capacity for wheat seed to a height of 8·6 m. Seed is to be scooped at the boot of the elevator and centrifugally discharged. Design the elevator for belt speed of 2·0 m/s.*

Solution. Given, Q_t = 10 tph, H_e = 8·6 m with centrifugal discharge, v_b = 2·0 m/s. Assume the coefficient of bucket filling, C_f = 0·85 and bulk density of wheat seed as 0·82 t/m^3. A belt with widely spaced buckets is to be employed for handling of wheat seed (Table 12.1). Selecting a three-ply belt weighing w_{be} = 1·5 kg/m, the drive pulley diameter is,

D_p = 125 × 3 = 375 mm say, 0·40m and r = 0·20 m

Pulley speed required is,

$$n = \frac{60 \times 2}{(0 \cdot 40\ \pi)} = 95 \text{ rpm}$$

Then $$h = \frac{895}{(95)^2} = 0{\cdot}10 \text{ m}$$

Since, $h < r$, a centrifugal discharge is confirmed from the provided bucket elevator.

$$R_p = \frac{h}{r} = \frac{0 \cdot 10}{0 \cdot 20} = 0{\cdot}5$$

Also $$D_p = \frac{2\, R_p\, V_p{}^2}{g} = \frac{2 \times 0{\cdot}5\,(2)^2}{9{\cdot}81} = 0{\cdot}41 \text{ m is recommended}$$

then D_p = 0·40 m is alright

Now $$\frac{V}{B_s} = \frac{Q_t}{3{\cdot}6\, V_b\, \rho\, C_f} = \frac{10}{3{\cdot}6 \times 2{\cdot}0{\cdot}82 \times 0{\cdot}75} = 2{\cdot}26 \text{ lit/m}$$

Deep buckets with V/B_s = 2·0 lit/m with 0·6 lit volume and 0·32 m pitch and 160 mm bucket width can be selected from Table 12.2 for this purpose weighing w_{bu} = 0·90 kg and belt width of 200 mm will be required. Then,

$$C_f = \frac{10 \times 0 \cdot 32}{(3 \cdot 6 \times 2 \times 0 \cdot 82 \times 0 \cdot 6)} = 0{\cdot}90$$

In this case, the bucket filling efficiency is improved over the assumed value of 0·85.

Now $H_c = H_e - 2\, r$ = (8·6 – 2 × 0·2) = 8·20 m

$$w_m = \frac{10}{(3 \cdot 6 \times 2)} = 1{\cdot}39 \text{ kg/m}$$

and $$w_b = \left(w_{be} + \frac{w_{bu}}{B_s}\right) = \left(1{\cdot}50 + \frac{0{\cdot}90}{0{\cdot}32}\right) = 4{\cdot}31 \text{ kg/m}$$

Then with C_s = 3 and equations (12.15) to (12.18)

$P_s = 3 \times 1{\cdot}39 \times 9{\cdot}81 = 40{\cdot}91$ N

and $P_2 = P_1 + 40{\cdot}91$

$P_3 = (P_1 + 40{\cdot}91) + 9{\cdot}81\ (1{\cdot}39 \times 8{\cdot}6 + 4{\cdot}31 \times 8{\cdot}2) = P_1 + 504{\cdot}88$

$P_4 = P_1 + (4{\cdot}31 \times 9{\cdot}81 \times 8{\cdot}2) = P_1 + 346{\cdot}70$

Considering m = 0·10 for high humidity operation and a = 180° or p radians,

$P_3 = P_4\, e^{0{\cdot}1p} = 1{\cdot}37\ (P_1 + 346{\cdot}70)$

Then, equating right hand sides for P_3 gives,

$P_1 + 504{\cdot}88 = 1{\cdot}37\ (P_1 + 346{\cdot}70)$

or $P_1 = 80{\cdot}81$ N

and $P_2 = 80{\cdot}81 + 40{\cdot}91 = 121{\cdot}72$ N

$P_3 = 80{\cdot}81 + 504{\cdot}88 = 585{\cdot}69$ N

$P_4 = 80{\cdot}81 + 346{\cdot}70 = 427{\cdot}51$ N

Therefore, from equation (11.24),

$$P_{ds} = \frac{(585{\cdot}69 - 427{\cdot}51) \times 2{\cdot}0}{986} = 0{\cdot}32 \text{ kW}$$

and from equation (11.25),

$$P_m = \frac{0{\cdot}32}{(0{\cdot}80 \times 0{\cdot}85)} = 0{\cdot}47 \text{ kW}$$

Hence, provide 0·75 kW (1.0 hp) electric motor for the elevator.

Example 12.2 : *If the elevator in Example 12.1 is supposed to deliver the grain by gravity discharge, determine the dimensions and power requirement. Assume weight of the belt and bucket as 2·1 kg/m and 1·75 kg, respectively.*

Solution. Assuming low velocity of 0·8 m/s, for gravity discharge pulley diameter is,

$D_p = 0{\cdot}6\ (0{\cdot}8)^2 = 0{\cdot}38$ m say, $D_p = 0{\cdot}40$ m and $r = 0{\cdot}20$ m

For this pulley, the speed required will be,

$$n = \frac{60 \times 0{\cdot}8}{(0{\cdot}4\ \pi)} = 38 \text{ rpm}$$

then $h = \dfrac{895}{(38)^2} = 0{\cdot}62$ m

and $R_p = \dfrac{h}{r} = \dfrac{0{\cdot}62}{0{\cdot}20} = 3{\cdot}1$

Since $R_p > 3{\cdot}0$, there will be a gravity discharge through this bucket elevator.

Now, the volume/pitch ratio required for bucket is,

$$\frac{V}{B_s} = \frac{10}{(3 \cdot 6 \times 0 \cdot 8 \times 0 \cdot 82 \times 0 \cdot 85)} = 4.98 \text{ lit/m}$$

Selecting a bucket with V/B_s = 5 lit/m which is widely spaced with 250 mm bucket width and 400 mm pitch with 300 mm wide belt, then,

$$w_m = \frac{10}{(3 \cdot 6 \times 0 \cdot 8)} = 3{\cdot}47 \text{ kg/m}$$

$$w_o = 2 \cdot 1 + \left(\frac{1 \cdot 75}{0 \cdot 4}\right) = 6{\cdot}47 \text{ kg/m}$$

and $P_s = 3 \times 3{\cdot}47 \times 9{\cdot}81 = 102{\cdot}12$ N

then $P_2 = P_1 + 102{\cdot}12$

$P_3 = (P_1 + 102{\cdot}12) + 9{\cdot}81\ (3{\cdot}47 \times 8{\cdot}6 + 6{\cdot}47 \times 8{\cdot}2) = P_1 + 915{\cdot}33$

$P_4 = P_1 + (6{\cdot}47 \times 9{\cdot}81 \times 8{\cdot}2) = P_1 + 520{\cdot}46$

also $P_3 = 1{\cdot}37\ (P_1 + 520{\cdot}46)$.

Therefore, $P_1 = 546{\cdot}75$ N, $P_2 = 648{\cdot}87$ N, $P_3 = 1462{\cdot}08$ N and $P_4 = 1067{\cdot}21$ N

An effective pull on head pulley will be,

$$P = 1462{\cdot}08 - 1067{\cdot}21 = 394{\cdot}87 \text{ N}$$

then $$P_{ds} = \frac{394 \cdot 87 \times 0 \cdot 8}{986} = 0{\cdot}40 \text{ kW}$$

Considering the transmission and motor efficiency of 0.80 and 0.85, respectively, provide 0·75 kW (1.0 hp) electric motor.

Example 12.3 : *Determine the head pulley diameters and their speeds for centrifugal, gravity and combined discharge of the granular material if the respective belt speeds are 2·0, 0·75 and 1·10 m/s.*

Solution. Considering the ranges of coefficients from equations (12.6, 12.8 and 12.9), the head pulley diameters will be,

(a) Centrifugal discharge

$$D_p = 0{\cdot}2\ (2)^2 = 0{\cdot}80 \text{ m}$$

and $$n = \frac{60 \times 2}{(0 \cdot 8\ \pi)} = 47{\cdot}74 \quad \text{say, 48 rpm}$$

also $$h = \frac{895}{(48)^2} = 0{\cdot}39 \text{ m}$$

and $$R_p = \frac{h}{r} = \frac{0 \cdot 39}{0 \cdot 40} = 0{\cdot}97$$

that is, $R_p < 1{\cdot}0$, which gives the centrifugal discharge.

(*b*) *Gravity discharge*

$$D_p = 0{\cdot}6\ (0{\cdot}75)^2 = 0{\cdot}34 \quad \text{say, } 0{\cdot}35\ \text{m}$$

and $n = 41$ rpm, $n = 0{\cdot}53$ m, $R_p = 3{\cdot}03$

which confirms the gravity discharge since $R_p > 3{\cdot}0$.

(*c*) *Combined discharge*

$$D_p = 0{\cdot}5\ (1{\cdot}1)^2 = 0{\cdot}605\ \text{m} \quad \text{say, } 0{\cdot}60\ \text{m}$$

then, $n = 35$ rpm, $h = 0{\cdot}73$ m and $R_p = 2{\cdot}43$. This is between 1·5 and 3·0 which gives moderate speed elevator with combined unloading of the material.

Nomenclature

B : width of the bucket, mm

B_s : bucket spacing on belt, m

C_s : coefficient of scooping grain in bucket, kg/kg

C_f : coefficient of bucket filling

D_p : diameter of drive or head pulley, mm

F : radial centrifugal force, N

g : gravitational constant, m/s^2

H : height of the bucket, mm

H_c : center to center distance between head and boot pulleys, m

H_e : total height of elevating the load, m

h : polar distance or height of pole, m

L : length of the bucket, mm

m : mass of grain in bucket, kg/m

N : number of plies in belt

n : head pulley speed, rpm

P : constant gravity force on grain in bucket, N

P_s : scooping pull (resistance) by the material during bucket filling, N

P_t : total pull (tension) on head pulley, N

P_1 : lowest pull (tension) in belt on tight side of boot pulley, N

P_2 : pull (tension) in belt on slack side of boot pulley, N

P_3 : pull (tension) in belt on slack side of head pulley, N

P_4 : pull (tension) in belt on tight side of head pulley, N

Q_t : capacity of bucket elevator, t/h

R : resultant force on grain in bucket, N

R_p : ratio of polar distance to the effective head pulley radius

r	: radius at which the particle moves around the center of head pulley, m
r_a	: radius at which the particle moves around the outer edge of bucket, m
r_p	: radius of the head pulley, m
t	: time of bucket movement around head pulley, s
V	: useful geometrical volume of bucket, lit
v_a	: absolute velocity of the particle, m/s
v_b	: velocity of the belt, m/s
v_c	: circumferential velocity of the particle, m/s
v_m	: velocity of the material moving in bucket, m/s
v_p	: circumferential velocity of the head pulley, m/s
v_s	: velocity of the sliding particle, m/s
W_{bu}	: total weight of buckets, kg
w_b	: weight of buckets and belt, kg/m
w_{be}	: weight of the belt, kg/m
w_m	: weight of material on belt, kg/m
x, y	: coordinates of the particle movement
α	: wrapping angle of belt on head pulley, deg
β	: angle of rotation of bucket from vertical position, deg
μ	: coefficient of friction between belt and head pulley surfaces
ρ	: bulk density of material to be conveyed, t/m^3
ω	: angular velocity of head pulley, rad/s

13

Screw Conveyers

13.1 Introduction

Screw conveyers are the oldest devices for carrying the bulk materials and consists primarily of conveyer screw rotating in a stationary casing which may be a trough or pipe. The material fed in the casing moves along its length by the rotation of screw which is supported by bearings at ends or intermediate positions too. The principle by which the material tends to move or slide is similar to the manner in which a nut travelled receives a translating motion when the screw fitted into it is turned. The load is prevented from turning with the screw by the weight and friction of the material against the casing walls. The material is discharged at the end of the casing or along the run, through openings in it. Intermediate openings have gates which are operated as and when needed.

Screw conveyers employed in seed processing plants, flour mills, chemical and building material manufacturing industries for transportation of various kinds of materials like pulverized, granular or sometimes small sized lumps. However, some processing operations such as heating, cooling, agitation, mixing, blending, extrusion, juice extraction etc. can be performed during conveying the dry and fluid ingredients enroute in food industries. It is the best feeder used for the other conveyers, particularly, where precise feed rates are required. A varying pitch screw can be employed for conveying the materials required to change the degree of their packing and also for controlling the discharge.

13.2 Types of screw conveyers

There are three types of screw conveyers, namely, horizontal, vertical and conveying screw pipes. They may be slightly inclined up to 20° with horizontal under first group or steeply inclined under the second group. The three types although have certain features common, they differ in other features and principle of operation. Screw conveyers are usually employed for relatively short distances up to 40 m in horizontal and up to 30 m heights in vertical types.

The advantages of screw conveyers are that they are compact and can be very easily adapted to the conjusted locations. They are simple in the construction, easy in maintenance and best suited for the controlled flow of material as feeders in the processing operations. They can be effectively sealed to prevent dust, hot or odorous loads and flame escaping or dust and dirt entering and can be jacketed to serve as the dryer or cooler during conveying. Their limitations are that they are not suitable to move wet, sticky and highly abrasive materials. They need a uniform feeding because overloads may cause clogging at intermediate bearings and can stop the entire screw. Besides this, the screw conveyers have a high power consumption, excessive wear of screw and casing surfaces with a substantial crushing of the material.

13.3 Horizontal screw conveyer

A helical flight screw is mounted on a shaft which is supported on bearings at the ends or even at intermediate places if screw is long (Fig. 13.1). The whole assembly is mounted in a trough for horizontal or slightly inclined screw conveyers up to 20° with horizontal. The trough may be semicylindrical or trapezoidal (Fig. 13.1) which receives material from the top opening at the screw starting end or at any intermediate places. With the rotation of a screw, the material is propelled towards the outlet(s) at the bottom of the trough below which the discharge spouts are mounted. The screws are usually a single leading, sometimes two or three leading for special purpose and may be right or left handed. Different types of screw flights are shown in Fig. 13.2. It may be solid and continuous with short pitch employed for dry granular or powdered materials (Fig. 13.1). A ribbon-like screw is also continuous but only with the band used for conveying lumpy and clinging materials. The flights for these types are made from the stampings or strips of steel sheet welded to the sheet helically. A paddle flight screw [Fig. 13.2 (*b*)] is made of blades fastened by unit bolts or pins on shaft and are used for the handling of slumping materials or those which have a packing tendency. A cut flight screw is also used for these materials which has a continuous flight but with out provided at the edges [Fig. 13.2 (*c*)]. For free flowing materials, pitch of the screw is taken equal to screw diameter and for slow flowing materials it is usually taken equal to 0·8 times the screw diameter. Shafts are tubular for easy connection and light weight with minimum diameter and length. Solid shafts are also in use. It is mounted on bearings such that the later should not interrupt the conveying of material. In order to take axial tensile force, the thrust bearing is provided at the discharge end of the screws. A shaft is driven by an electric motor or prime mover through transmission gear reduction unit or worm reducer.

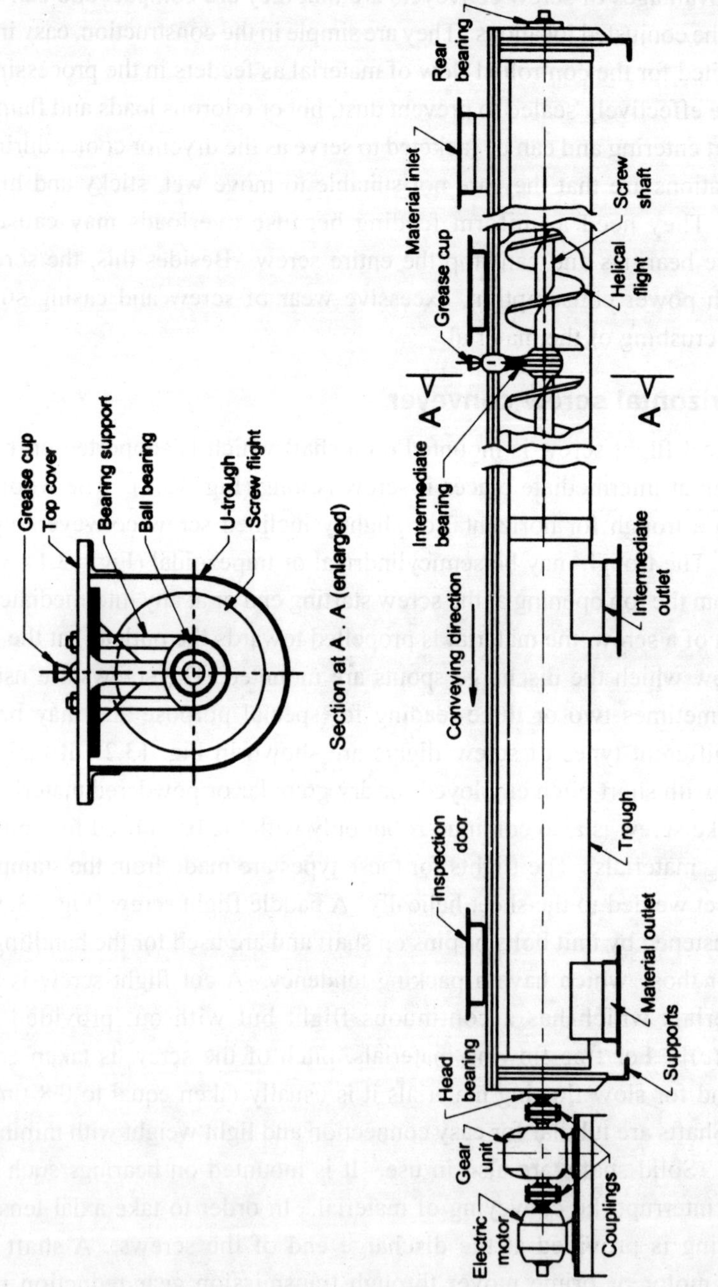

Fig. 13.1 *Screw conveyer assembly*

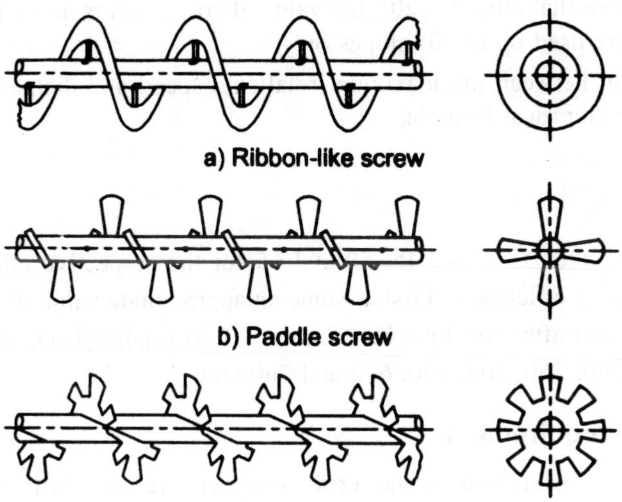

a) Ribbon-like screw

b) Paddle screw

c) Cut flight screw

Fig. 13.2 *Types of screw flights*

13.3.1 Conveying capacity

The capacity of screw conveyer depends on the screw diameter, mean conveying velocity of the material, specific weight of the material and the loading efficiency taken as 0·125 for slow flowing highly abrasive material, 0·25 for slow flowing light abrasive materials, 0·30 for free flowing light abrasive material and 0·40 for free flowing non-abrasive materials. Capacity of the screw conveyer is given by,

$$Q_t = \left(\frac{\pi}{4}\right) D_s^2\, v_m\, \rho\, \varepsilon\, 3600$$

or, $$Q_t = 2827\, D_s^2\, v_m\, \rho\, \varepsilon \quad (13.1)$$

and $$v_m = \frac{S n}{60} \quad (13.2)$$

However, the approximate capacity is also given by,

$$Q_t = \left(\frac{60\pi D_s^2}{4}\right) S n \rho \varepsilon$$

$$= 47\, D_s^2\, S\, n\, \rho\, \varepsilon \quad (13.3)$$

The conveying capacity is reduced with an increase in the angle of inclination for the inclined screw with ascending motion of the material. For this, equations (13.1) and (13.3) should be multiplied by a correction factor C_s depending on the slope as below.

β, deg	:	0	5	10	15	20	25
C_s	:	1·00	0·90	0·81	0·73	0·65	0·52

It can be seen that after $\beta = 20°$, the value of correction factor is reduced fastly and hence, screws are used up to 20° slopes only.

The relation between the maximum rotational speed and diameter of the screw is expressed by an empirical formula,

$$n_{max} = \frac{C}{D_s^{0.5}} \tag{13.4}$$

Coefficient taken C as 30, 45, 50 and 65 for the respective above conditions of material for loading efficiency. First, assume an approximate value of screw diameter in equation (13.4) and after checking by equation (13.3) finalize from standard values *i.e.* 100, 125, 150, 200, 250, 300, 450, 600 and 760 mm.

13.3.2 Power requirement

The movement of material in the screw conveyer is resisted by the friction on trough and screw surfaces, internal resistance of the material friction within itself and that offered by the bearing friction. The weight of material per unit length of the conveyer is given by

$$\dot{w}_m = \frac{Q_t}{(3.6\, v_m)} \tag{13.5}$$

Length of material rubbing on helical screw surface per revolution is given as,

$$L_s = \frac{3\pi D_s}{4\cos\beta} \tag{13.6}$$

The power requirements for friction with trough, screw surface, within material and bearings are given as below,

$$P_t = \frac{g w_m L_c \mu_t v_t}{986} \tag{13.7}$$

$$P_s = \frac{g w_m L_c n \mu_s \mu_t}{59176} \tag{13.8}$$

$$P_m = (P_t + P_s)\,\mu_m \tag{13.9}$$

$$P_b = \frac{g w_c L_b \mu_b \pi d n}{59176} \tag{13.10}$$

The total power requirement at the drive shaft is given by the sum of above powers multiplied by the factor for initial acceleration of the material, which yields,

$$P_{ds} = f_a (P_t + P_s + P_m + P_b) \tag{13.11}$$

The value of factor f_a ranges between 1·10 and 1·25.

For ascending inclined screw conveyers, an additional power required is,

$$P_i = \frac{w_m H_{V_t}}{986} \tag{13.12}$$

The electric motor power requirement in each case is obtained by,

$$P_{mo} = \frac{P_{ds}}{\eta_t \, \eta_m} \tag{13.13}$$

The transmission efficiency is taken 0·80 for spur gears and 0·95 for fully lubricated gears and bearings.

13.4 Vertical screw conveyer

It is also called as the screw lift which consists of a vertical shaft with a continuous helical flight suspended on a thrust bearing at the top and plain bearing at bottom rotating inside the tubular casing (Fig. 13.3). Usually, the material is fed with the help of a short horizontal screw and is delivered through a pipe connected at the top end of the casing. Bottom portion of casing has a decreasing diameter upward or the screw pitch is smaller. Intermediate bearings are also provided to the shaft, if height of conveyer is large. A steep slope screw conveyer also falls under this group. The material received at the bottom of a vertical screw is rotated and the centrifugal forces tend to move upward along a helical path in axial direction. Vertical screw conveyers are employed for transportation of pulverized, granular and fine-sized lumpy materials. They have a limited capacity and lifting up to a height of 15 m or rarely up to 30 m. Their advantages and drawbacks are similar to those of horizontal screw conveyers. A special case of application of a vertical screw conveyers is the post hole digger. The soil is dug with a drilling apex and the role of casing is performed by cylindrical walls of borehole.

Critical speed of screw

In case of horizontal conveyers the screw speed is limited by the highest value but in the vertical screw there should be a sufficient speed (critical) to develop a centrifugal force so as to move the material upward axially. Considering the forces exerted on moving particle as shown in Fig. 13.4. A centrifugal force and a tangential force on casing surface are given as;

$$F_c = \frac{mv^2}{R} \tag{13.14}$$

and
$$F_t = \mu_c F_c = \mu_c \frac{mv^2}{R} \tag{13.15}$$

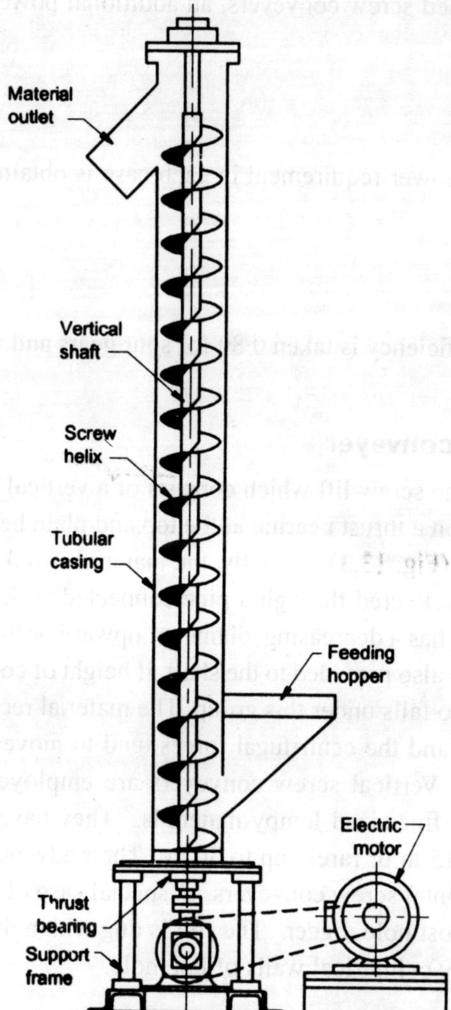

Fig. 13.3 *Vertical screw conveyer*

This force has two components acting on the helix surface; ($\mu_c m v^2 \cos\alpha / R$) and ($\mu_c \, \mu_s \, m \, v^2 \sin\alpha / R$). Then, from force balance (Fig.13.4),

$$\tan(90 - \alpha - \phi_c) = \frac{P}{F_t} = \frac{m g R}{\mu_c m v^2}$$

or $$\cot(\alpha + \phi_c) = \frac{gR}{\mu_c v^2}$$

then $$v^2 = \frac{gR}{\mu_c \cot(\alpha + \phi_c)}$$

$$= \frac{g\,R\tan(\alpha+\phi_c)}{\mu_c}$$

or $$v_c = \left[\frac{g\,R\tan(\alpha+\phi_c)}{\mu_c}\right]^{0\cdot5} \qquad (13.16)$$

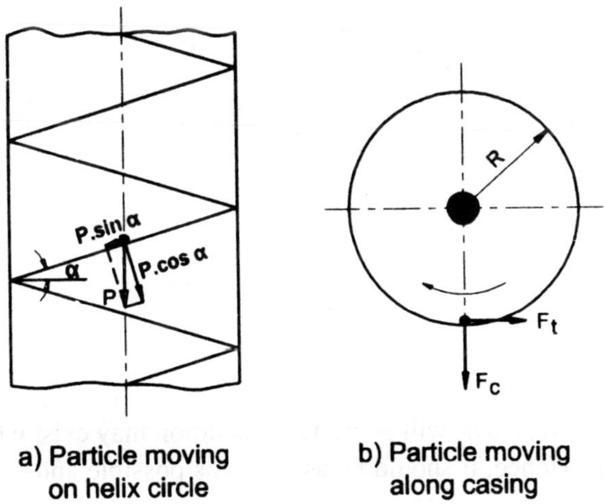

Fig. 13.4 *Forces on particle moving in vertical screw conveyer*

Putting, $v_c = \pi\, D\, n_c/60$ in equation (13.16), critical rotational speed of the screw is given by the following relationship.

$$n_c = \frac{30}{\pi}\left[\frac{2g}{D\,\mu_c}\tan(\alpha+\phi_c)\right]^{0\cdot5}$$

or $$n_c = 42\cdot30\left[\frac{\tan(\alpha+\phi_c)}{D_p\,\mu_c}\right]^{0\cdot5} \qquad (13.17)$$

The screw dimensions and surfaces of screw and casing should satisfy the following conditions so as to move the material in axial upward direction in vertical screw conveyers.

(*a*) Rotational speed of the screw should be higher than the critical speed to initiate and continue the vertical movement of a particle on screw flight surface. As the speed is a function of helix angle of flight α which has least value in equation (13.17), n_c is regarded as the lowest speed of screw rotation at which vertical conveying or self-emptying of material is just possible.

(*b*) An angle of helix must be higher than the friction angle *i.e.* $\alpha > \phi_c$, otherwise equation 13.17 will be an indeterminate.

(*c*) If $(\alpha + \phi_c) = 90^o$, then n_c will be ∞. Hence, efforts of the screw design should be such that $\alpha < (90 - \phi_c)$. Then,

$$\tan \alpha < \tan (90 - \phi_c)$$

$$< \cot \phi_c$$

$$< \frac{1}{\tan \phi_c}$$

$$< \frac{1}{\mu_c}$$

also $$\tan \alpha = \frac{S}{\pi D_p}$$

then $$\frac{S}{\pi D_p} < \frac{1}{\mu_c}$$

or $$\frac{S}{D_p} < \frac{\pi}{\mu_c} \tag{13.18}$$

(*d*) If $\mu_c \Rightarrow 0$, then also n_c will be ∞. This condition may exist when the casing inside surface is polished. Hence, it should be as rough as possible and certainly not a smooth tube.

13.5 Conveying screw pipe

It consists of a helical attached to the inside surface of pipe which is mounted on two pairs of support rollers. The bulk material fed at one end is advanced due to rotation of pipe and pouring over the helix walls and discharged at the other end. For effective motion of the material, the pipe is usually mounted with a slight slope towards the discharge end. They are employed for granular materials only in different processing operations like drying, intermixing, roasting etc. over a length of 20-25 m depending on the residential period required for a particular process operation. Conveying screw pipes have simple design, high reliability, isolation from surrounding over a wide range of conveying lengths and capacities. However, they have large dimensions and are structurally gothic. Equation (13.2) gives the conveying velocity and capacity of screw pipes is determined in a similar way as that for the horizontal screw conveyers by using the equations (13.1) and (13.3). The pipe length for a given processing time t is determined by the formula,

$$L_p = v_m \, t \tag{13.19}$$

The screw pitch is taken as 0·5 D_p and filling coefficient as 0·2 to 0·3. Pouring action of material will take place when gravitational force exceeds the frictional force on pipe surface. The speed of screw rotation is given by the force balance as,

$$n = 42 \cdot 3 \left[\frac{\sin(\alpha - \phi_c)}{(D_p \sin\phi_c)} \right]^{0 \cdot 5} \tag{13.20}$$

The maximum speed of rotation which is a critical speed for not to rotate the granular material along with the pipe surface can be obtained by the substitution of $\alpha = 90° + \phi_c$ as below,

$$n_c = 42 \cdot 3 \ [D_p \sin \phi_c]^{-0 \cdot 5}. \tag{13.21}$$

The centrifugal force F will be equal to the gravity force on material at $\alpha = 180°$ *i.e.* at the top point of the pipe.

Then the maximum speed will be

$$n_m = 42 \cdot 3 \ D_p^{-0 \cdot 5} \tag{13.22}$$

which is higher than critical speed n_c, usually, taken as 0·50 to 0·65 times n_m.

Power requirement

Power in screw pipes is required for two functions *viz.*, convey of material by lifting within helical path and resistance offered by rollers during rotation. The total power requirement at drive shaft is given by the formula,

$$P_{ds} = \frac{2\pi ng}{986} \left[\frac{w_m D_p}{4} + \frac{(w_m + w_c)}{\cos\beta_p} \cdot \frac{D_{pr}}{2} \cdot \frac{(2\mu_p + \mu_r)d}{D_r} \right] \tag{13.23}$$

An electric motor power requirement can be obtained by using equation (13.13).

Example 13.1 : *A screw conveyer of 20 t/h capacity is required in pulse milling industry for transportation of pigeonpea grain with bulk density 850 kg/m³ to a horizontal distance of 16·5 m. An angle of material friction with screw and trough surfaces is 19° and angle of internal friction of grain is 23·5°. If the screw speed is limited to 33 rpm, determine the screw diameter and mean particle velocity in the conveying direction. Assuming screw assembly weight as 12 kg/m, bearing friction coefficient as 0·025, shaft diameter 50 mm and transmission efficiency as 75%, determine the horse power requirement at drive shaft and that of the prime mover with 80% efficiency.*

Solution. Given, $Q_t = 20$ t/h, $\rho = 850$ kg/m³ $= 0 \cdot 85$ t/m³, $L_c = 16 \cdot 5$ m, $n = 33$ rpm, assume $\alpha = 15°$.

For granular low abrasive material

$$\varepsilon = 0 \cdot 40 \text{ and } S = 0 \cdot 8 \ D$$

From equation (13.3),

$$Q_t = 47 \ D_s^2 \ S \ n \ \rho \ \varepsilon = 47 \ D_s^2 \ (0 \cdot 8 \ D_s) \ n \ \rho \ \varepsilon$$

or $$D_s^3 = \frac{Q_t}{47 \times 0 \cdot 8 n\rho\varepsilon} = \frac{20}{47 \times 0 \cdot 8 \times 33 \times 0 \cdot 85 \times 0 \cdot 4} = 0 \cdot 047$$

then $$D_s = 0 \cdot 36 \text{ m or } 360 \text{ mm}$$

From equation (13.1),

$$v_m = \frac{20}{2827\,(0.36)^2 \times 0.85 \times 0.4} = 0.16 \text{ m/s}$$

Power requirement

Given $w_c = 12$ kg/m, $\mu_b = 0{\cdot}025$, $d = 0{\cdot}05$ m, $\eta_t = 0{\cdot}75$, $\eta_m = 0{\cdot}80$,

$$\mu_s = \mu_t = \tan(19) = 0{\cdot}344,\ \mu_m = \tan(23{\cdot}5) = 0{\cdot}435$$

also $$w_m = \frac{20}{(3.6 \times 0.16)} = 34{\cdot}72 \text{ kg/m}$$

and $$L_s = \frac{3\pi \times 0.36}{(4\cos 15)} = 0{\cdot}878 \text{ m/rev}$$

Power requirement for resistance of different forces will be,

$$P_t = \frac{34.72 \times 9.81 \times 16.5 \times 0.344 \times 0.16}{986} \times 0.746 = 0{\cdot}3135 \text{ kW}$$

$$P_s = \frac{34.72 \times 9.81 \times 16.5 \times 0.878 \times 33 \times 0.344 \times 0.344}{59176} = 0{\cdot}3256 \text{ kW}$$

$$P_m = (0{\cdot}3135 + 0{\cdot}3256) \times 0{\cdot}435 = 0{\cdot}2780 \text{ kW}$$

$$P_b = \frac{12 \times 9.81 \times 16.5 \times 0.025\pi \times 0.05 \times 33}{59176} = 0{\cdot}0043 \text{ kW}$$

Assuming initial acceleration factor as 1.20 for pigeonpea grain, total drive shaft power from equation (13.11) will be,

$$P_{ds} = 1{\cdot}20\ (0{\cdot}3135 + 0{\cdot}3526 + 0{\cdot}2780 + 0{\cdot}0043) = 1{\cdot}1057 \text{ kW}$$

and $$P_{mo} = \frac{1.1057}{0.75 \times 0.80} = 1{\cdot}84 \text{ kW}$$

Hence, provide 2·24 kW (3.0 hp) electric motor for driving the screw.

Example 13.2 : *A seed treater has 18 t/h capacity and fed through a vertical screw conveyer with 300 mm pitch and diameter rotating at 150 rpm. The seed has 18° angle of friction with casing surface and the helix angle of moving particle with horizontal is 30°. Check whether the material will move' in vertical direction, if not, suggest suitable modifications.*

Solution. Given, $Q_t = 18$ tph, $S = D_p = 0{\cdot}30$ m, $\rho = 150$ rpm, $\rho = 0{\cdot}80$ t/m^3, $\phi_c = 18°$ and hence,

$$\mu_c = \tan 18° = 0{\cdot}325,\ \beta = 30°$$

An angle of flight is,

$$\alpha = \tan^{-1}\left(\frac{1}{\pi}\right) = 17{\cdot}65°$$

But $\alpha > \phi_c$ is an essential requirement for moving the material in vertical direction. Therefore, providing $\alpha = 20°$, $\tan \alpha = 0{\cdot}364$, screw flight pitch required is,

$$S = 0{\cdot}30 \times 0{\cdot}364\ \pi = 0{\cdot}34 \text{ m}$$

Also, $\left(\frac{S}{D}\right) < \left(\frac{\pi}{\mu_c}\right)$ *i.e.*, $\left(\frac{0 \cdot 34}{0 \cdot 30}\right) < \left(\frac{\pi}{0 \cdot 325}\right)$ or, $1 \cdot 133 < 9 \cdot 66$

and $n_c = 42 \cdot 30\,(0 \cdot 30 \times \sin 18)^{-0{\cdot}5} = 138{\cdot}93$ rpm

Thus, in a given screw the pitch should be increased from 0·30 to 0·34 m and then the vertical motion of material will take place inside the screw since μ_c is not equal to zero.

Nomenclature

C : coefficient for determination of screw diameter

C_s : correction factor for capacity reduction in screw conveyers on slope

D_p : diameter of screw pipe, m

D_{pr} : pipe ring diameter driven on rollers, mm

D_r : diameter of roller, mm

D_s : diameter of screw, m

d : diameter of screw shaft, mm

F : centrifugal force exerted on material, N

F_c : centrifugal force developed due to particle velocity, N

F_t : tangential force on casing surface, N

f_a : factor considering initial acceleration of material

g : gravitational constant, m/s^2

H : total vertical height of conveying the material, m

L_b : length of the screw conveyer from head-end to tail-end bearings, m

L_c : length of conveying the material, m

L_p : length of screw conveyer pipe, m

L_s : length of friction on helical screw surface per revolution, m

m : mass of grain to be conveyed, kg

n : speed of the screw, rpm

n_c : critical speed of screw, rpm

n_m : maximum speed of screw, rpm

P : constant gravity force due to grain, N

P_b : Power requirement for bearing friction, kW

P_{ds} : Power requirement at drive shaft, kW

P_i : Power requirement for ascending inclined conveyance of material, kW

P_m : Power requirement for friction within material, kW
P_{mo} : Power requirement for electric motor, kW
P_p : constant gravity force due to the pipe loads, N
P_s : Power requirement for friction of material on screw surface, kW
P_t : Power requirement for friction of material on trough surface, kW
Q_t : capacity of screw conveyer, t/h
R : radius of head pulley, mm
S : pitch of the screw, mm
t : residential time of material in screw conveyer, s
v : peripheral speed of the screw, m/s
v_c : critical peripheral speed of the screw, m/s
v_m : mean conveying velocity of material in screw conveyer, m/s
v_t : velocity of material in contact with trough surface, m/s
w_c : weight of screw conveyer assembly, kg/m
w_m : weight of material per unit length of screw conveyer, kg/m
α : angle of inclination of screw flight, deg
β : angle of inclination of screw conveyer with ascending motion of material, deg
β_p : half the angle of inclination of screw pipe, deg
η_{mo} : efficiency of an electric motor, %
η_t : efficiency of transmission system, %
ϕ_c : angle of material friction on casing surface, deg
μ_b : coefficient of friction for bearings
μ_c : coefficient of friction of material on casing surface
μ_m : coefficient of friction of material itself within the grain
μ_p : coefficient of friction of pipe ring
μ_r : coefficient of friction of roller axle
μ_s : coefficient of friction of material on screw surface
μ_t : coefficient of friction of material on trough surface
ρ : specific weight of the material to be conveyed, t/m^3
ε : loading efficiency factor

14

Pneumatic Conveyers

14.1 Introduction

Pneumatic conveyer moves the granular bulk loads through a closed duct/pipeline in air stream caused due to the pressure difference between two ends of duct produced by a blower or vacuum (exhaust) fan. Granular or powdered material to be conveyed is mixed with the stream of air and moves forward in suspended state and partially sliding on the bottom of the duct. The capacity of pneumatic conveyer may be to the tune of 100 tph over a short to longer distance up to hundreds of meters. They are employed in seed processing plants, cotton ginning plants and *dal* mills for conveying seed, foodgrains, milled material, seed cotton etc. Pneumatic conveyers are at advantages due to the tight conveyance of material hence, it is not exposed to the ambient conditions, without any loss of material, easy connections in any direction with simple joints are possible. Conveyance of the material from multi-points to a single point and vice versa is possible through the pneumatic conveyers. Some of the processing unit operations *e.g.* cooling, drying etc. can be combined with this system. However, the pneumatic system has the limitations that it requires high specific power with heavy wear of the duct and other parts due to the material flow. Moist or sticky product cannot be transported with this system and the particle size less than 10 mm is very difficult to separate from the air stream.

14.2 Conveying systems

According to the pressure difference in a duct, pneumatic conveyers are classified in three groups, namely, suction or vacuum systems [Fig. 14.1 (*a*)], pressure systems [Fig. 14.1 (*b*)] and the combination of the two. Suction systems operate below atmospheric pressure and the material is introduced through a suction nozzle into the duct. In case of pressure system, the material is conveyed in a stream of compressed air *i.e.* pressure above atmospheric due to which the pressure systems are more efficient as compared with the suction systems to convey the granular material over longer distances and heights.

The mechanical components of the system are more clustered at the end of the suction duct but they are at the beginning of the duct in case of pressure systems. In combined type systems, the blower and mechanical components are provided in the mid of the conveying path.

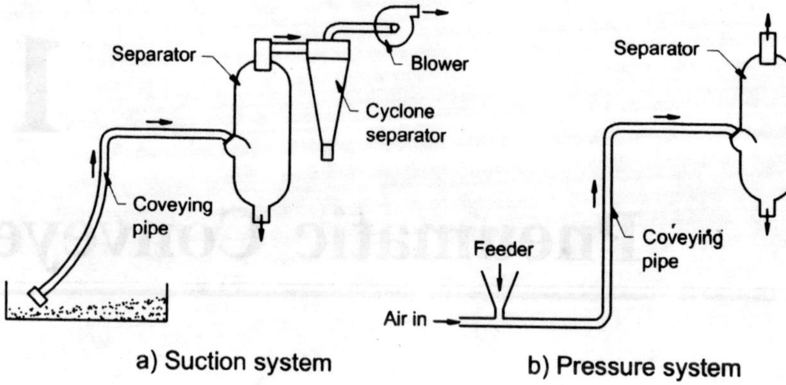

Fig. 14.1 *Types of pneumatic conveying systems*

In a suction system, the granular material is taken up by the suction pipe and conveyed through the duct to a separator where the material is separated from air stream and falls to the separator bottom due to velocity fall of air stream. Air is then sucked in a dust collecting cyclone, passes through the blower and ejected to the atmosphere. The system may have 2 to 4 suction pipes with flexible hoses. A low-pressure pneumatic conveyer (pressure 0·2 to 0·3 MPa) uses high air velocity and low density air powered by a centrifugal fan. High-pressure systems use low velocity and high density air (pressure 0·3 to 0·4 MPa) created by a compressor which moves to the closed reservoir feeder where the material is taken-up by the air stream and conveyed through the duct. The discharge of pressure systems may be single or multiple at atmospheric pressure. In a combined suction-pressure system, the material is taken-up by the suction pipe/nozzle into the duct, enters the separator at intermediate point in a path and further conveyed to the pressure system. A blower at the intermediate point sucks in the air from the duct and discharges it to the next duct at the end of which a separator is provided for separation of material from the air stream. The suction system is best suited for unloading granular materials where the point of unloading may move or there are number of locations from which the material is to be taken-up. Unloading of trailers, trucks or wagons are the examples. The pressure system is to be used where the point of discharge varies *e.g.* loading of trucks or storage bins.

14.3 System devices

The main devices required for a pneumatic system are suction pipe/nozzle, feeders, conveying duct, with fittings, blowers, separators and filters.

Feeders : Feeding of material to be conveyed to the air stream is important owing to the effect/damage to the material, power required for feeding and economics of the feeder. The material may be directly sucked-up through a flexible hose [Fig. 14.1 (*a*)], it may be introduced directly into the fan [Fig. 14.1 (*b*)] or may be metered into the air stream by a bucket wheel, screw auger, injector or column feeder (Fig. 14.2). The simplest, most efficient and cheapest feed is directly into the fan, but some materials may be damaged due to this. The suction hose system is convenient, but hose velocity must be high. The bucket wheel and auger meters provide a uniform material flow without any damage, but are expensive and may have back air leakage through the meter. However, this loss is nonsignificant. Column meter is simple, but back leakage may be significant at low column heights.

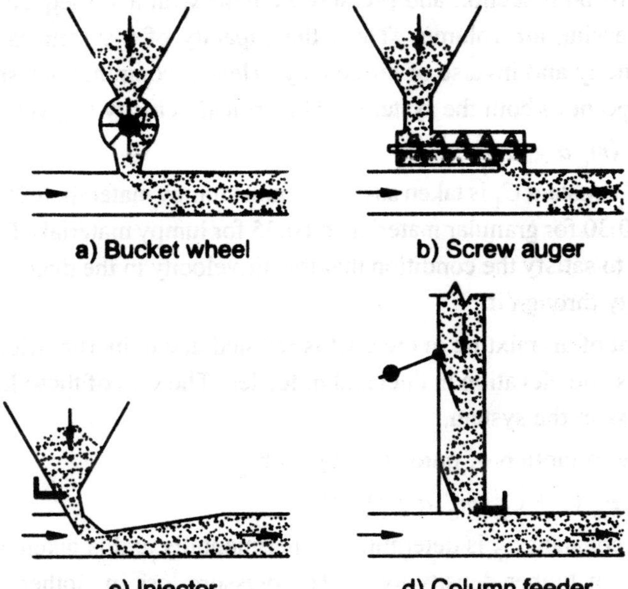

Fig. 14.2 *Feeders for pneumatic systems*

Separators : The conveyed bulk load is separated from air stream at the point where discharge is required, either at the end or the intermediate point of the system. A separator is essentially a vessel of larger cross section than the duct so that air velocity falls as it enters the separator and material also fall down the bottom of the separator. Air stream changes the direction and forming a cyclone at the centre of vessel, discharged from the vertical pipe provided at the top of the vessel. The major portion is separated at the bottom and finer particles are separated inside the central one.

Air blowers : Pneumatic conveyers usually employ centrifugal fans for operation of the pressure systems and turbo-blowers (5000 to 7000 rpm) for suction system.

14.4 Conveyer calculation

In order to determine the pressure drop and horse power requirement of pneumatic system, necessary data needed is the capacity of the system, length of the duct, its configuration and material properties. In suction systems, the mass concentration of the mixture $\leq 1{\cdot}0$ or up to 5 for high vacuum. In pressure systems, is equal to 8 to 25 depending on conditions and 60 to 150 for air-activated material flow. Value of mass concentration depends on the looseness and lumpiness of the material, lump formation, moisture content, nature of duct such as turns, lifts and activation by air etc. Too low concentrations are not recommended due to increasing specific power requirement with decreasing concentrations. For lumpy material, the duct diameter should be greater than or equal to three times the size of largest lump. The air velocity increases from initial point to the end point of both suction and pressure systems with a subsequent decrease in air density and increasing air volume. Thus, the capacity of a system is the function of square of air velocity and inverse of air density. Hence, computations should be carried out at the initial point of both the systems. The critical velocity is given by,

$$v_c = C_1 (m_c\, a\, g\, D_c)^{0{\cdot}5} \tag{14.1}$$

The empirical constant C_1 is taken as 0·1 for moist lumpy material, 0·25 for complicated path of the duct; 0·30 for granular material and 0.35 for lumpy material. The duct diameter must be checked to satisfy the condition that the air velocity in the duct is greater than the critical air velocity through it.

The movement of air mixture in the duct is resisted due to inertia, friction against duct wall, length, turns and elevation in duct and in feeder. The sum of these losses in pressure gives the total loss in the system.

The resistance to motion of pure air is given by,

$$p_a = 0{\cdot}1\, p_m\, (1 + C_2\, m_c\, a\, g\, D_c/v^2) \tag{14.2}$$

An empirical constant C_2 is determined experimentally and is taken as 0·01 to 0·075, the lower values for higher density loads. The pressure loss in isothermal air motion is computed by the following equation.

$$p_m = 0\cdot 10\left[\left(\frac{m_a^2\, R T_a\, \lambda\, L_c}{A^2 g\, D_c (10)^8} + \frac{p_e^2}{100}\right)^{0\cdot 5} - 1\right] \tag{14.3}$$

The universal gas constant $R = 29{\cdot}3$. The coefficient of resistance is taken as 0·016 to 0·02 for 150 mm ducts, 0·015 to 0·018 for 175 mm ducts and 0·014 to 0·016 for 200 mm duct diameter. The resolved conveying length of duct includes, straight length and equivalent lengths of turns (Table 14.1) and the pressure at the end of the duct is taken as 0·105 MPa. An inertia loss associated with the acceleration of load to the working speed is given by,

$$p_i = 0{\cdot}5\, v^2\, \rho_o\, (1 + \beta\, m_c)\, (10)^{-5} \tag{14.4}$$

The index of relative velocity of particles to be conveyed is taken as 0·35 to 0·85 for granular and 0·60 to 0·85 for powdered materials. Pressure loss in short rising duct at a height h_t is given by,

$$p_h = \rho_o g h_t (1 + \mu) (10)^{-3} \tag{14.5}$$

Table 14.1 : Equivalent length (m) of duct turns for 0·7 to 1·0 m radius

Angle of turn, deg	*Equivalent length for type of material to be conveyed*		
	Pulverized	*Granular*	*Lumpy*
30	4	5	15
45	6	8	25
90	10	15	45

Considering the loss of pressure in feeder, the total pressure required in a system is,

$$p_t = p_a + p_i + p_h + p_f \tag{14.6}$$

Air flow rate at the discharge of blower is,

$$V_a = k_1 V_c$$

The coefficient of power leakage in the duct, feeder etc. is taken as 1·10 to 1·15. Blower electric motor power is given by,

$$P_m = \frac{k_2 W_b V_a}{986 \eta_m} \tag{14.7}$$

The coefficient of conveyance is taken as 0·8 to 0·9, the blower efficiency is taken as 0·65 to 0·85 and the theoretical work of blower per unit volume of sucked in air in isothermal compression which is given by,

$$W_b = 2303 \, p_e \log \left(\frac{p_a}{p_e} \right) \tag{14.8}$$

Nomenclature

A : cross section area of conveying duct, m^2

a : size of the largest lump, m

C_1, C_2 : imperial constants

D_c : diameter of conveying duct, m^2

g : gravitational constant, m/s^2

h_t : height of a short rising duct, m

k_1 : coefficient of power leakage in the duct

k_2 : coefficient of conveyance

L : conveying length of duct, m

L_c : resolved conveying length of duct, m

m_a : mass flow rate of air, kg/h

m_c : mass concentration of mixture, kg/ m^3

P_m : power requirement for blower electric motor, kW

p_a : resistance to the motion of pure air, N/mm^2

p_e : pressure at end of the duct, N/mm^2

p_f : pressure loss in feeder, N/mm^2

p_h : pressure loss in short rising duct, N/mm^2

p_i : pressure loss due to inertia of the material, N/mm^2

p_m : pressure loss in isothermal air motion, N/mm^2

p_t : total pressure required in the conveying system, N/mm^2

Q_t : capacity of pneumatic conveyer, t/h

R : universal gas constant (= 29.3)

T_a : absolute ambient temperature, °C

V_a : volumetric air flow rate at blower discharge, m^3/s

V_c : volumetric air flow rate required for conveying, m^3/s

V_t : volumetric capacity of pneumatic conveyer, m^3/s

v : air velocity through duct, m/s

v_c : critical air velocity through duct, m/s

W_b : theoretical work of blower per unit volume of sucked air in isothermal compression, Nm/m^3

β : index of relative velocity of particles to be conveyed

η_m : efficiency of the electric motor, %

ρ_o : initial bulk density of material to be conveyed, t/m^3

λ : coefficient of resistance to air flow

15

Optimization

15.1 Introduction

Many design problems have n number of potential solutions. The role of designer is to choose the best one through the precise compromise and using the optimization techniques. It is necessary to optimize the process where several hundred variables work together with complex interactions by dividing it into more sizable number of variables or unit operations for the greatest benefit. The optimization helps two fold: one, to operate the equipment and systems at the maximum efficiency and two, to save the use of costly material in fabrication. Moreover, it can also reduce the time and energy inputs to the process, ultimately, leading to the enhancement of economy which is the aim of any enterprise.

15.2 Steps in optimization

In order to get the maximum profit, the designer has to divide the objectives such as minimum operating cost and investment, maximum efficiency and yield, reduction in labour, energy and time and also the repairs and maintenance costs etc. The first step in optimization will be the intuitive judgement of the designer for section of the sub-objective because for instance minimising cost per unit of production will not necessarily maximise the profit per unit time due to market factors such as quality and delivery. Secondly, it is necessary to select the objective function (what to optimize ?) which may be related to input cost, material consumption, surface area etc. with the affecting variables/ factors. Thirdly, the relationship is to be derived between the objective function and the influencing variables in the form of mathematical model such that the number of variables exceeding the number of mathematical equations. From these relationships the values of the variables are determined which give the optimum value (maximum or minimum) of the objective function.

15.3 Methods of optimization

Various methods or techniques are used for the optimization of solution for a particular problem namely, analytical, search, linear, dynamic programming and case study etc.

15.3.1 Analytical method

The objective function which can be expressed in terms of mathematical relationship can be differentiated and set equal to zero to determine the maximum or minimum. For multiple variable function, the partial derivatives are obtained with respect to each variable and a set of simultaneous equations is obtained such that the number of equations is equal to the number of variables. Using the Lagrange multipliers, the equations and variables are made identical and the optimum solution is obtained. It should be noted that for this method, the function must be in the form of the differentiable mathematical equation. Many design problems are of the iterative type and are solved by using numerical method or by trial and error. The analytical method gives only one solution and no any direction or path to reach it.

One-variable function

The maximum or minimum (extremes) of a function can exist only when the first derivative is equal to zero. If the second derivative of a function is negative, a maximum exists, if it is positive, a minimum exists.

When dealing with the multi-variable problem and there are equality constraints on the variables, use of the Lagrange multipliers in analytical approach is the powerful tool for getting the optimum solution. First step in this method is to develop the Lagrange expression and secondly, to determine the values of independent variables subject to the given constraints of the problem and also to optimize the objective function simultaneously. In this method the first rule is that the number of Lagrange multipliers to be introduced must be equal to the number of constraining equations and second rule is that the Lagrange expression must be equal to the objective function plus the product of Lagrange multiplier and constraint. The later must be in the form of equation set equal to zero.

Let x, y,.... be the independent variables, u is the objective function and w is the Lagrange expression with λ_1, λ_2,..... are the Lagrange multipliers. Then the objective function is,

$$u\ (x,\ y, \ldots\ldots)$$

and constraining equations are,

$$v_1\ (x,\ y, \ldots) = 0 \text{ and } v_2\ (x,\ y,\ \ldots) = 0 \text{ etc}$$

Lagrange expression is,

$$w\ (x,\ y, \ldots\ldots) = u\ (x,\ y,\ \ldots\ldots) + \lambda_1 v_1\ (x,\ y,\ \ldots\ldots) + \lambda_2 v_2\ (x,\ y, \ldots\ldots)$$

After the development of Lagrange expression λ, its partial derivatives are obtained with each variable and set equal to zero for an extreme. These n number of equations for n variables are simultaneously solved to obtain the extreme conditions. Since, each constraining equation introduces one additional equation, one Lagrange multiplier for equation is introduced to compensate for this.

15.3.2 Search method

When the problem is such that the analytical method cannot be used due to the nature of mathematical relationship and constraints, the search method is used. In case of single-variable problem, where the objective function is unimodal, it is simply required to compute the values of objective function at uniform intervals of independent variables until the desired extreme value is obtained. Although not very efficient, this method requires less time for simple problems. Search by golden section is an efficient method to obtain the extreme value of the objective function. Two-variable problem can be plotted in the form of contours (Fig. 15.1) which is sliced through the three-dimensional model of the function.

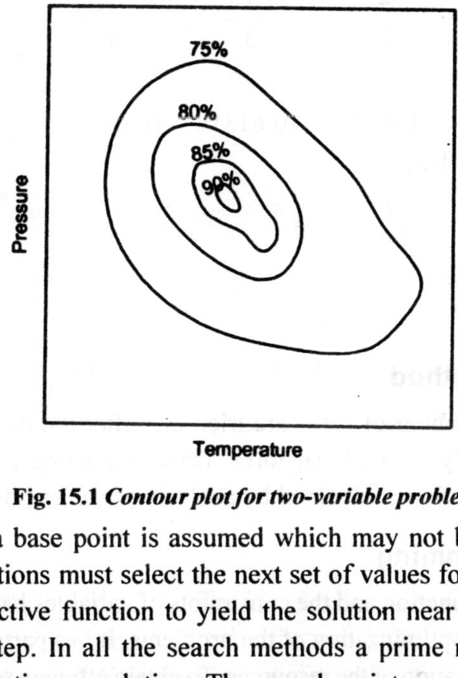

Fig. 15.1 ***Contour plot for two-variable problem***

In this method, a base point is assumed which may not be the optimum solution. Then this set of conditions must select the next set of values for the independent factors and evaluate the objective function to yield the solution near to the optimum solution through every such step. In all the search methods a prime requirement is that there should be only one optimum solution. The search points are symmetrically placed in search by golden section method from both ends of the search region. Let, this distance be L_1. On the basis of these two calculations a portion of the search region, where optimum solution does not occur, can be eliminated by examining the two results. A system which has only one peak point, called unimodal, is considered here. Let, in the search region 0 to 1, A and B are its two ends and C and D be the two points at distance L_1 from A and B, respectively in the specified region, such that the search value of C is higher than that of D. According to the modality concept, the peak cannot lie in the region A to D and hence, this region can be eliminated for further search. Thus, now the search region is from D to B which is equal to the length L of the specified region and point C is at a distance $(1 - L_1)$

from point *B*. The characteristic of L_1 is such that L_1^2 from *B* is the point *C* only which is again equal to $(1 - L_1)$. From this,

$$L_1^2 = 1 - L_1$$

or $$L_1^2 + L_1 - 1 = 0$$

then $$L_1 = \frac{-1 \pm [1^2 + 4 \times 1(1)^{0.5}]}{2} = 0{\cdot}618$$

Thus, for every pair of new points the search region is reduced by 0·618. If m be the number of steps, N be the number of total calculations and I_R be the region of isolation, then the following gives the efficiency of this method.

Step number	1	2	3	...m
New calculation	2	1	1	... 1
Total calculation	2	3	4	... N = 1 + m
Region of isolation of optimum	0·618	$(0{\cdot}618)^2$	$(0{\cdot}618)^3$	... $I_R = (0{\cdot}618)^m$

From this, it follows that,

$$I_R = (0{\cdot}618)^{(N-1)}$$

and $$N = 1 + \frac{\log I_R}{\log 0{\cdot}618}$$

15.3.3 Case study method

In this method the number of cases are tried one after the other and the results are presented either graphically or in tabular form. Thus, evaluating various solutions to the problem, the optimum solution is obtained by analysis of the graphs and tables.

15.3.4 Linear programming

When the objective function and the constraints of variables have linear relationship, this technique is used for optiminzation of the problems. It has particularly application in deciding the optimum utilization of the resources. To obtain efficient solution of the resource problem of linear programming the algorithms have been developed and the SIMPLEX algorithm is widely used for this purpose.

Example 15.1: *For the minimum total cost of the refined oil storage tank determine the diameter/ height ratio, if the bottom and top cost 0·85 and 1·37 times the cost of vertical wall. Also compute the tank dimensions for 450 m^3 capacity.*

Solution. Let, C_b, C_t, C_s and C_T be the costs of bottom, top, side, total tank, respectively. Also *V*, *D* and *H* is the respective volume, diameter and height of the oil tank. The total cost of the tank is given by,

$$C_T = (\pi D H)\, C_s + \left(\frac{\pi}{4}\right) D^2 (C_b + C_t) \quad (1)$$

and $$V = \left(\frac{\pi}{4}\right) D^2 H, \text{ or } H = \frac{4V}{\pi D^2}$$

Substituting the value of H in equation (1), gives,

$$C_T = \left(\frac{4V}{D}\right) C_s + \left(\frac{\pi}{4}\right) D^2 (C_b + C_t) \quad (2)$$

Differentiating equation (2) with respect to D and equating to zero,

$$\frac{d\, C_T}{d\, D} = -\frac{4V}{D^2} \cdot C_s + \frac{\pi D}{2} (C_b + C_t) = 0$$

or $$D^3 = \frac{8V}{\pi} \cdot \frac{C_s}{C_b + C_t} \quad (3)$$

Again differentiating equation (3) yields,

$$\frac{d^2\, C_T}{d\, D^2} = \frac{8V\, C_s}{D^3} \quad (4)$$

which is the positive term, hence, the function has minimum. The *D*/*H* ratio optimum value for minimum cost of the tank is given by substituting for *V* in equation (3) as below,

$$D^3 = \frac{8[(\pi/4) D^2\, H]}{\pi} \cdot \frac{C_s}{C_b + C_t}$$

or $$\frac{D}{H} = \frac{2 C_s}{C_b + C_t}$$

Substituting bottom and top costs in terms of side wall,

$$\frac{D}{H} = \frac{2 C_s}{(0{\cdot}85 C_s + 1{\cdot}37 C_s)} = 0{\cdot}90$$

or $H = 1{\cdot}11\, D$

The volume of tank is given by

$(\pi/4)\, D^2\, H = 450$

or $(\pi/4)\, D^2\, (1{\cdot}11\, D) = 450$

then $D^3 = 516{\cdot}18$

and $D = 8{\cdot}02$

also $H = 1{\cdot}11 \times 8{\cdot}02 = 8{\cdot}90$ m

Example 15.2: *An oil extraction unit requires to store 200 t raw oil to be filtered having the density of 0·926 t/m³. Determine the dimensions of required rectangular tank in order to have the minimum sheet required for fabrication having the following objective and constant functions, respectively,*

$$u\,(L, B, H) = 2LB + 2BH + HL \tag{1}$$

and $$v\,(L, B, H) = LBH - V \tag{2}$$

where, L, B, H and V are the length, breath, height and volume of the required tank, respectively.

Solution. Introducing the Lagrange multipler in equation (1) and (2), the Lagrange expression will be,

$$w\,(L, B, H) = 2LB + 2BH + HL + \lambda\,(LBH - V) \tag{3}$$

Differentiating the Lagrange expression (3) partially with respect to each variable and equation to zero yield the following expressions.

$$\frac{\delta w}{\delta L} = 2\,B + H + \lambda\,B\,H = 0 \tag{4}$$

$$\frac{\delta w}{\delta B} = 2\,L + 2\,H + \lambda LH = 0 \tag{5}$$

$$\frac{\delta w}{\delta H} = 2\,B + L + \lambda\,LB = 0 \tag{6}$$

$$\frac{\delta w}{\delta \lambda} = L\,B\,H - V = 0 \tag{7}$$

From equation (4),

$$\lambda = -\frac{(2B + H)}{BH}$$

Inserting this value in equation (5) gives,

$$2L + 2H - \left[\frac{(2B + H)}{BH}\right]LH = 0$$

Multiplying the above expression by $B\,H$,

$$2\,L\,B\,H + 2\,B\,H^2 - 2\,B\,L\,H\; - LH^2 = 0$$

or $\quad 2\,B = L$ or $\qquad B = 0{\cdot}5\,L$

Also from equation (6),

$$\lambda = -\frac{(2B + L)}{LB}$$

Inserting this value of λ in equation (4) yields,

$$2B + H - \frac{(2B + L)}{LB} \text{ BH} = 0$$

Multiplying this expression by $L\,B$ gives,

$$2\,L\,B^2 + L\,B\,H\ - 2\,B^2\,H - L\,B\,H = 0$$

then $\quad L = H$

Now $\quad V = L\,B\,H\ = L \times 0{\cdot}5\,L \times L = 0{\cdot}5\,L^3$

and volume of oil is $\left(\frac{200}{0{\cdot}926}\right) = 216\ \text{m}^3$

so that, $0{\cdot}5\,L^3 = 216$

and $L = 7{\cdot}56$ m, $B = 3{\cdot}78$ and $H = 7{\cdot}56$ m

Example 15.3: *A self-emptying conical bottom 50 t capacity storage bin is to be provided for a roller mill to supply wheat grain (bulk density 788 kg/m³) at controlled rate. Developing the objective function and constraints, determine the bin dimensions. Assume the hopper bottom angle as 45° with horizontal and per unit area cost of conical bottom and flat top as 1·20 and 0·75 times the cost of side cylindrical wall.*

Solution. For the minimum cost of bin the objective function is,

$$u\,(D, H) = (\pi\,D\,H)\,C_s + (0{\cdot}25\,\pi\,D^2)\,C_t + \left[0{\cdot}5\pi\,D\left(\frac{D}{2\tan\alpha}\right)\right]C_b \tag{1}$$

and constraining equation is,

$$u\,(D, H) = (0{\cdot}25\,\pi\,D^2)\,H + (0{\cdot}25\,\pi\,D^2)\left(\frac{h}{3}\right) - V \tag{2}$$

where, H and D are the height and diameter of storage bin, h is the vertical height of cone. C_s, C_t and C_b are the unit costs of side cylinderical wall, flat top and hopper bottom, respecitvely, V is the volume of the bin. The half cone angle $\alpha = 45°$ and $h = D/(2\sin\alpha)$. Then equation (1) and (2) will be,

$$u\,(D, H) = (\pi\,DH)\,C_s + (0{\cdot}25\,\pi D^2)\,(0{\cdot}75\,C_s) + \left[0{\cdot}5\pi D\left(\frac{D}{2\tan\alpha}\right)\right](1{\cdot}20\,C_s)$$

$$= (\pi\,D\,H)\,C_s + (0{\cdot}25\,\pi\,D^2)\,(0{\cdot}75\,C_s) + \left(\frac{\pi\,D^2}{2{\cdot}83}\right)(1{\cdot}20\,C_s)$$

$$= \pi\, C_s\, [D\, H + 0{\cdot}19\, D^2 + 0{\cdot}42\, D^2]$$

$$= \pi\, C_s\, (D\, H + 0{\cdot}61\, D^2) \quad (3)$$

$$v\,(D, H) \quad = (0{\cdot}25\, \pi\, D^2\, H + (0{\cdot}25\, \pi\, D^2)\cdot \frac{D}{(3 \times 2 \tan 45)} - V$$

$$= (0{\cdot}25\, \pi\, D^2 H) + \left(\frac{\pi\, D^3}{24}\right) - V \quad (4)$$

From equation (3) and (4) and Lagrange multiplier, the Lagrange expression is,

$$w\,(D, H) = \pi\, C_s\, (D\, H + 0{\cdot}61\, D^2) + \lambda \left[(0\cdot 25\pi\, D^2\, H) + \left(\frac{\pi\, D^3}{24}\right) - V\right] \quad (5)$$

The partial derivation of equation (5) yields the following expressions equating to zero with D, H and λ :

$$\frac{\partial w}{\partial D} = \pi\, C_s\, (H + 1{\cdot}22\, D) + \lambda \left[0\cdot 5\pi\, D\, H \frac{\pi\, D^2}{8}\right] = 0 \quad (6)$$

$$\frac{\partial w}{\partial H} = \pi\, C_s\, D + \lambda\, (0{\cdot}25\, \pi\, D^2) = 0 \quad (7)$$

$$\frac{\partial w}{\partial H} = (0{\cdot}25\, \pi\, D^2 H + \left(\frac{\pi\, D^3}{24}\right) - V = 0 \quad (8)$$

From equation (7),

$$\lambda = \quad \frac{-4\, C_s}{D}$$

Substituting this value in equation (6) yields,

$$\pi\, C_s\, (H + 1{\cdot}22\, D) - \left(\frac{4\, C_s}{D}\right)\left[0{\cdot}5\pi\, D\, H + \frac{\pi\, D^2}{8}\right] = 0$$

or $\quad H + 1{\cdot}22\, D - 2\, H - 0{\cdot}5\, D \quad = 0$

or $\quad 0{\cdot}72\, D - H = 0$

$$H \; = 0{\cdot}72\, D$$

Volume of grain $= \dfrac{50}{0\cdot 788} = 63{\cdot}45\ \text{m}^3$

Substituting the value of H, the volume from equation (8) will be,

$$V = (0{\cdot}25\ \pi\ D^2)\ (0{\cdot}72\ D) + \left(\frac{\pi D^3}{24}\right) = 63{\cdot}45$$

then $0{\cdot}696\ D^3 = 63{\cdot}45$

or $D = 4{\cdot}50$ m

then $H = 4{\cdot}5 \times 0{\cdot}72 = 3{\cdot}24$ m

and $h = \dfrac{D}{(2\tan 45)} = \dfrac{4\cdot 5}{2\times 1} = 2{\cdot}25$ m

Example 15.4: *Soybean oil with 8·83 (10)$^{-3}$ MN/m^3 specific weight is to be stored in 40 m^3 tank which rests on a concrete platform. Tank is to be covered by a sheet roof. The cylindrical wall thickness is a function of oil pressure on it. Assuming the allowable stress in wall as 78 MN/m^2 and unit costs of tank bottom with concrete, cover and sides as Rs. 800, 360 and 80 t, respectively, determine tank size, wall thickness and total cost of the tank.*

Solution. Let, the cost of bottom with concrete, cover and side wall be C_b, C_t and C_s, respectively and C_T be the total cost of the tank of diameter and height (H). Then,

$$C_T = (0{\cdot}25\ \pi\ D^2)\ (C_b + C_t) + \pi\ D\ H\ C_s \tag{1}$$

Substituting the costs,

$$C_T = (0{\cdot}25\ \pi\ D^2)\ (800 + 360) + \pi\ D\ H\ (80\ t)$$

or $$C_T = 911\ D^2 + 251{\cdot}33\ t\ D\ H \tag{2}$$

Volume of the tank (V) is given by,

$$V = 0{\cdot}25\ \pi\ D^2\ H \tag{3}$$

and the hydrostatic pressure (p) on side wall due to oil is,

$$p = H\rho_f$$

or $p = 8{\cdot}83\ (10)^{-3}\ H$

Now, side wall thickness is given by

$$t = \frac{pD(10)^3}{2f} = \frac{8\cdot 83(10)^{-3}\, D\, H(10)^3}{2\times 78}$$

or $$t = 0{\cdot}0566 \tag{4}$$

From equation (3),

$$DH = \frac{V}{0\cdot 25\pi D} = \frac{40}{0\cdot 25\pi\, D} = 50{\cdot}93\ D$$

Substituting this value in equation (4),

$$t = 0{\cdot}0566 \left(\frac{50{\cdot}93}{D}\right)$$

or $$t = \frac{2{\cdot}883}{D} \tag{5}$$

then $$D = \frac{2{\cdot}883}{t}$$

and $$D^2 = \frac{8{\cdot}312}{t^2} \tag{6}$$

also $$D\,H = 17{\cdot}67 \tag{7}$$

Substituting equation (6) and (7) in equation (2) gives,

$$C_T = 911 \left[\frac{8{\cdot}312}{t^2}\right] + 251{\cdot}33\,[17{\cdot}67\,t]$$

or $$C_T = \frac{7572{\cdot}2}{t^2} + 4441\,t \tag{8}$$

Let, the region of search be $t = 0$ to 10 mm

t, mm	C_T, Rs.	Remarks
6	160086	Since the cost is increasing with t, it should
7	217764	be < 6 mm, but < 5 mm is not permissible
5	111328	$t = 5$ mm is optimum value

From equation (5),

$$D = \frac{2{\cdot}883}{5} = 0{\cdot}58 \text{ m}$$

and from equation (7),

$$H = \frac{17{\cdot}67 \times 5}{0{\cdot}58} = 152{\cdot}3 \text{ m}$$

Then $$\frac{H}{D} = \frac{152{\cdot}3}{0{\cdot}58} = 262{\cdot}6$$

These dimensions of the tank are not practicable and the H/D ratio is too high. Hence, the optimization may be examined with respect to D. From equation (7), $D\,H = 15{\cdot}87 \times 5 = 79{\cdot}35$. For less than this value $t = 5$ mm is valid. In this case $C_s = 80 \times 5 = 400$ is fixed. Analysing equation (2) again for D and substituting the value of H from equation (3).

$$H = \frac{V}{0 \cdot 25\pi D^2} = \frac{40}{0 \cdot 25\pi D^2} = \frac{50 \cdot 93}{D^2}$$

Then $C_T = 0{\cdot}25\ \pi D^2 (800 + 360) + \pi D \left(\frac{50 \cdot 93}{D^2}\right) (400)$

or $$C_T = 911\ D^2 + \frac{64001}{D} \quad (9)$$

Let, the region of search be $D = 0$ to 5 m. Then from equation (9),

D, m	C_T Rs.	
4·0	30576	
3·5	29446	D should be < 3·5 m
3·0	29533	
3·6	29585	D should be between 3·5 and 3·0 m
3·4	29355	D < 3·4 m
3·3	29315	OK
3·2	29329	D > 3·2 m

For $D = 3{\cdot}3$ m $\quad C_T =$ Rs. 29315

and $$H = \frac{50 \cdot 93}{(3 \cdot 3)^2} = 4{\cdot}68 \text{ m}$$

also $$\frac{H}{D} = \frac{4 \cdot 68}{3 \cdot 30} = 1{\cdot}418$$

Thus, it can be concluded that the wall thickness is not the controlling parameter of the minimum cost but it is the tank diameter.

Nomenclature

B : breadth of rectangular tank, m

C_b : cost of the tank bottom, Rs

C_s : cost of the tank sides, Rs

C_T : total cost of the tank, Rs

C_t : cost of the tank top cover, Rs

D : diameter of circular tank, m

H : height of the tank, m

h : vertical height of conical bottom of the tank, m

I_R : region of isolation

L : length of the rectangular tank, m

L_1 : distance of search points from both the ends of search region

m : number of steps in search method

N : number of total calculations in each step of the search method

p : hydrostatic pressure exerted on tank walls, MN/m^2

t : tank wall thickness, mm

u, v, w : functions

V : volume of the tank, m^3

α : half cone angle of conical bottom, deg

λ : Lagrange multiplier

ρ_f : density of fluid to be stored in tank, kg/m^3

Appendices

Appendix A : Standard steel plate dimensions (mm)

Thickness	:	5, 5.5, 6, 7, 8, 9, 10, 11, 12, 14, 16, 18, 20, 22, 25, 28, 32, 36, 40, 45, 50, 56, 63, 71, 80
Width	:	160, 180, 200, 220, 250, 280, 320, 355, 400, 450, 500, 560, 630, 710, 800, 900, 1000, 1100, 1250, 1500, 1600, 1800, 2000, 2200, 2500
Length	:	6000 to 10000, 1000 interval

Appendix B: Nominal outside diameters (mm) for process vessels

100, 125, 150, 200, 250, 300, 350*, 400, 500, 600, 700, 800, 900, 1000, 1100, 1200, 1300*, 1400, 1500*, 1600, 1700*, 1800, 1900*, 2000, 2100*, 2200, 2300*, 2400, 2600, 2800, 3000, 3200, 3400*, 3600, 3800*, 4000, 4250*, 4500, 4750*, 5000

**These values should be used when unmarked sizes are either too small or too large.*

Appendix C: Specific weight and Poisson's ratio for different metals

Material	Specific weight, N/m^3	Poisson's ratio
Steel	$7{\cdot}70\ (10)^4$	0·30
Copper	$8{\cdot}79\ (10)^4$	0·35
Brass	$8{\cdot}35\ (10)^4$	0·35
Aluminium	$2{\cdot}65\ (10)^4$	0·34
Iron	$7{\cdot}74\ (10)^4$	0·28

Appendix D: Modulus of elasticity (E, 10^3 N/mm^2) for different metals

(*a*) *Ferrous materials*

Material	Design temperature, °C						
	0	20	100	200	300	400	500
Carbon steel							
C < 0.3%	196	196	195	190	182	170	—
C > 0.3%	211	210	207	199	190	173	—
Co-Mo steels and Cr steels (up to 3%)	211	210	207	201	194	184	172
High carbon steels	193	193	190	186	180	173	160

(*b*) *Nickel and nickel alloys*

Material	Design temperature, °C						
	200	300	400	500	600	700	800
Nickel	211	204	188	165	140	117	109
Nickel copper alloy (70%, 30%)	188	180	176	169	162	155	150
Ni, Cr and ferrow alloy (75%, 15%, 10%)	218	207	200	176	160	130	119

(*c*) *Copper and copper alloys*

Material	Design temperature, °C						
	20	50	100	150	200	250	300
Copper (99·98% Cu)	112	111	110	100	106	104	101
Commercial brass (66% Cu 34% Zn)	98	97	96	95	91	89	86
Leaded tinbronze (88% Cu, 6% Sn, 1·5% Pb, 4·5 Zn)	91	90	89	87	84	82	78
Phosphor bronze (85·5% Cu, 12·5% Sn, 10% Zn)	105	103	102	98	95	91	85
Muntz (59% Cu, 39% Zn)	107	102	98	91	83	77	—
Cupro nickel (80% Cu, 20% Ni or 70% Cu, 30% Ni)	133	131	129	126	124	121	118

References

Arora, S.C. and S. Domkundwar (1984). A Text Book of Heat and Mass Transfer. Third Edition, Dhanpat Rai and Sons, Delhi.

Bhattacharyya, B.C. (1998). Chemical Equipment Design. CBS Publishers and Distributors, New Delhi.

Brown, G.G. (1995). Unit Operations. CBS Publishers and Distributors, New Delhi.

Chakravarty, A. (1985), Post Harvest Technology of Cereals, Pulses and Oilseeds. Oxford and IBH Publishing Company, New Delhi.

Charms, S.E. (1978). The Fundamentals of Food Engineering. Third Edition. The AVI Publishing Company, New York.

Coulson, J.M. and J.F. Richardson (1977). Chemical Engineering, Volume 4. Third Edition, Pergamon Press, New York.

Davande, S.D. (1999). Process Design of Equipments. Central Technopublications, Nagpur.

Earle, R.L. (1983). Unit Operations in Food Processing. Second Edition, Pergamon Press, New York.

Foust, (1982) Principles of Unit Operations, 2 Edition, John Wiley New York.

Groggins, (1970) Unit Operations in Organic Synthesis, 5th Edition, Mcgraw Hill Book Co. New York.

Heldman, D.R. (1975). Food Process Engineering. AVI Publishing Company, New York.

Joshi, M.V. and V.V. Mahajan (1996). Process Equipment Design. Third Edition, MacMillan India Ltd, New Delhi.

Kuprits, Y.N. (1967). Technology of Grain Processing and Provender Milling. Israel Program for Scientific Translations, Jerusalem.

Mujumdar, A. (1995). Handbook of Industrial Drying, Volume 1. Second Edition, Marcel Dekker Inc, New York.

McCabe, W.L. and J.C. Smith (1985). Unit Operations in Chemical Engineering. International Students Edition, McGraw Hill Book Company, New York.

Shukla, BD and Gyanendra Singh ((2004) Drying and Dryers, Jain Brothers, New Delhi

Spivakovsky, A.O. and V.K. Dyachkov (1985). Conveying Machines, Volume I and II. Mir Publishers, Moscow.

Published by Sunil Kumar Jain for Jain Brothers 16/873, East Park Road, Karol Bagh, New Delhi-110005 and typeset by *Niyati Graphics*, 754/23, Khajoor Road, Karol Bagh, New Delhi-110005 and printed at G. Print Process, Daya Basti, Delhi-110035.

Index

D

H

I

J

K

L

M